KU-334-769

CONTENTS

MAP USES, SCALES AND ACCURACIES FOR ENGINEERING AND
ASSOCIATED PURPOSES

by

The Committee on Cartographic Surveying

Foreword

A basic purpose of the Committee on Cartographic Surveying,
Surveying and Mapping Division, ASCE, as stated in the ASCE Official
Register, is, "to aid the map user in selecting the map he needs." To
this end, this Committee was authorized a Task Committee "for the
preparation of a manual on the selection of map types, scales, and
accuracies for engineering and planning" in 1970 (6). This special
publication has been prepared to provide assistance in communications
between engineers who use maps and those who produce maps. The
introduction will describe the proposed purpose and scope of the special
publication, provide background information on steps taken toward its
publication, and review the content of its chapters.

CHAPTER 1 - INTRODUCTION

By: Carl B. Feldscher, [1] Member, ASCE

Purpose and Scope of the Special Publication

Maps are a necessary tool for performing engineering functions on
many types of projects, large and small. "Obtainment and judicious use
of maps of the proper type, scale, accuracy, and quality are the keys to
success in such work . . ." (7)

What is a map? ASCE Manual and Report on Engineering
Practice No. 34 (2) contains the following definition: "map —
A representation on a plane surface, at an established scale of
the physical features (natural, artificial, or both) of a part
or the whole of the Earth's surface, by use of signs and
symbols, and with the method of orientation indicated. Also, a
similar representation of the heavenly bodies. A map may
emphasize, generalize, or omit the representation of certain
features to satisfy specific requirements. The type of
information which a map is designed primarily to convey is
frequently used, in adjective form, to distinguish it from maps
of other types. A map should contain a record of the
projection on which it is constructed."

---

[1] Hydrographic Engineer, Retired, Palm Harbor, FL 33563

1

As stated, maps are generally classified by their content. They are also classified by their primary use. Examples of maps classified in these ways are: assessment, bathymetric, cadastral, contour, engineering, forestry, geological, highway, hydrographic, hypsographic, hypsometric, index, isogonic, key, land classification, landscape, magnetic, nautical, reconnaissance, right-of-way, and zoning. These are defined in ASCE-M&R No. 34.

These maps are either planimetric or topographic. A planimetric map presents the horizontal positions only for features represented. A topographic map may show the same features as a planimetric map, but uses contours or other symbols to show mountains, valleys and plains, and in the case of hydrographic charts, uses symbols and numbers to show depths in bodies of water (2).

Furthermore, Ref (7) states: "For engineering purposes, maps are further classified by scale and accuracy, each map constituting a selective representation of the features and details important to the specific needs of their particular users. Moreover, it borders on the trite to say it is impossible to produce an all-purpose map. Scale limitations, delineation problems, cluttering and/or obscuration of details whenever too much is placed on a map, and differing requirements preclude both the feasibility and desirability of endeavoring to compile an all purpose map. Segregation and emphasis of specific details are often accomplished by symbolization and color delineation on transparent overlays. Each separate map should be compiled to fulfill its current and known and anticipated future uses."

How are maps of the required type, scale, accuracy, and quality selected? These requirements vary with the stage in performing the engineering function (such as reconnaissance and design), the type of project, the current and planned land use, and the character of the topography. Also, there is a necessary correlation between feasible and adequate map scales, contour intervals, and accuracies. Preparation of the special publication has been aligned towards assisting engineers in determining the map scale, contour interval, accuracy, and symbols which would be proper for a given project. The emphasis has been on large-scale maps - those at scales larger than 2000 feet per inch (1:24,000). The special publication presents recommendations for determination of map scales and contour intervals, based on experience and a study of past and current mapping practice; for determination of map accuracy, based on a proposed alternative to U.S. National Map Accuracy Standards for large-scale maps; and for determination of map symbols, based on a review of existing publications and procedures on map symbology related to large-scale mapping. Information on sources of existing maps, i.e., map availability, is also presented in the special publication since such maps are often very useful in some stages of a project. The special publication should provide practical approaches, not readily found in other publications, to the solution of many problems encountered in selecting and obtaining maps for engineering and associated purposes. It is not expected to provide a replacement for

2

existing map manuals and texts. Selected data on sources of information on the subjects of the special publication chapters are presented in a Bibliography.

Content of the Special Publication

The names of the chapters and their authors are as follows: Chapter 1 - Introduction, Carl B. Feldscher; Chapter 2 - Maps for Engineering and Associated Work by Stages, William T. Pryor; Chapter 3 - Selection of Maps for Engineering and Associated Work, William T. Pryor; Chapter 4 - Engineering Map Accuracy Standards and Testing, Dean C. Merchant; Chapter 5 - Map Content and Symbols, Robert P. Jacober, Jr.; Chapter 6 - Map Availability, Robert L. Brown; and Appendix A - Bibliography, Carl B. Feldscher.

The following review of the content of the succeeding chapters of the special publication is intended to provide an insight into the applications of this publication to: selecting maps, defining and testing map accuracy, selecting and using map symbols, and securing published or other available maps. It is a source of information about maps and mapping which should be of value to engineers. Updating of the contained information in future issues of Surveying Engineering and in new special publication editions is planned by the Committee.

In the second chapter, Maps for Engineering and Associated Work by Stages, the author identifies major fields of engineering and associated work for which maps are needed. Pryor points out that you should consider many factors in map selection, including: cost, accuracy, land use, the character of the topography, the nature of the engineering project, and the stage in accomplishing the work. For basic map types, designated by characteristics relating to their content and construction, he presents criteria governing the selection of map scale and contour interval and identifies maps used in such stages of the work as planning, design, location, and construction.

In the third chapter, Selection of Maps for Engineering and Associated Work, Pryor presents a means of determining, on the basis of established criteria and within a framework of practical values, the contour intervals and map scales best suited to stages of a specific project. In defining the framework of values, maps are categorized by their use in design or planning. For each use category there is a specific range of suitable and compatible values of contour interval and scale. The author has prepared charts and tables showing the relationship of these map parameters to selection criteria, such as map uses, project stages, land use, and topography. The tables are a necessary tool in the selection process which he has devised. Several examples provide assistance in understanding this process.

Pryor also in this chapter proposes an indirect method of determining whether or not ground surveys are needed to meet requirements for accuracy. The value of the ratio of the specified contour interval to the specified map scale is applied to find the expected adequacy of photogrammetric mapping. He introduces for map selection criteria the use of the term "Characterization" to indicate

3

the density of detail required on the map, and the term "Map-
Compilation Factor", to indicate the level of accuracy and reliability
expected to be obtained in photogrammetric mapping.

   In the fourth chapter, Engineering Map Accuracy Standards and
Testing, the author proposes an alternative to U.S. National Map
Accuracy Standards (USNMAS) for large-scale maps. Accuracy is specified
and tested in terms of limiting standard errors and the limiting
absolute errors, in keeping with accepted statistical procedures. He
defines the terms as they apply to maps. Measurements in coordinate
directions on the map are related to measurements on the ground. The
principles of Engineering Map Accuracy Standards (EMAS), applicable to
map specification and compliance testing, are presented.

   In an explanation of EMAS, Merchant shows the relationship of this
standard to the USNMAS. Appendices to the chapter include: a statement
of the USNMAS; a discussion of the selection of appropriate error values
or error tolerances, including suggestions by Pryor; examples of EMAS
specifications in SI units and in English units; and an example of
compliance testing.

   The fifth chapter, Map Content and Symbols, is based on research by
Jacober of current practice in the selection and use of map symbols in
large - scale mapping and of computer-assisted mapping. Benefits of the
use of standard symbols over the current use of a variety of symbols are
discussed (5). The primary emphasis in the chapter is on symbols for
engineering maps at scales between 20 feet per inch and 400 feet per
inch (between 1:240 and 1:4,800) and on two-color, such as black-and-
white, symbology. Tables of such symbols are presented. Most of these
symbols were selected from a large collection of legends and tables
while other necessary symbols were designed by the writer.

   These symbols did not duplicate those in other existing sets of
symbols referred to in the text, those produced by the American National
Standards Institute (ANSI) for large-scale maps and by government
agencies for smaller-scale maps and charts. To progress toward the
standardization of symbology, it is recommended that, on large-scale
engineering maps, features should be identified by symbols in the ANSI
set, by symbols in this special publication or, as appropriate, by
symbols in the sets produced by government agencies, such as the U.S.
Geological Survey, the National Ocean Survey, and the Defense Mapping
Agency. The author invites submission of proposed symbols for features
not identified in these sets to the Committee on Cartographic Surveying
for inclusion in future editions of this special publication.

   Other subjects discussed in this chapter include alternatives for
the depiction of existing and proposed features; conventions and
guidelines for map content, such as the placement of lettering, the
displacement of symbols, marginal information, map text, and titles; and
the results of a study of available computer software for graphics which
are presented through eight tables defining and comparing features and
applications of nine software systems. Jacober programmed each symbol
tabulated in this chapter to assure that they could be displayed using
the computer.

In the sixth chapter, Map Availability, Brown reviews and discusses
the wide range of maps and map by-products available and their
sources. Maps may be obtained from agencies of the Federal, state,
county, and municipal government and certain other agencies. In an
appendix to this chapter, there is, for each map producing agency of the
U.S. Government, a list by name and scale of maps and charts or
categories of maps and charts produced by that agency, a description of
the use or purpose of the product, and the addresses of offices of the
agency.

The addresses of state highway and geological survey (or
equivalent) departments for each of the 50 states and the District of
Columbia are listed in another appendix. These agencies are information
sources for state mapping. The information on map availability in this
chapter should assist in obtaining maps for many applications in
engineering and associated work.

Appendix A - Bibliography identifies readily available, published
source materials on subjects relating to those of the special
publication chapters.

Background

The Committee on Cartographic Surveying under the Chairmanship of
Edward H. Sokolowski made initial plans for the special publication in
1968. Its preparation was begun as a joint effort of the American
Society of Civil Engineers (ASCE), the American Congress on Surveying
and Mapping (ACSM), and the American Society of Photogrammetry (ASP).
At a meeting of representatives of these organizations in March 1970, it
was decided that the selection of maps for the following general
engineering functions would be considered: (1) transportation; (2) land
development; (3) urban planning; (4) drainage, sewers, and water supply;
(5) boundary surveys; (6) dams and reservoirs; (7) waterfront
facilities; and (8) mining. With some modifications, these functions
have been the basis for subsequent effort on the special publication
which has been coordinated through a Task Committee, formed in October
1970, with Antonio M. Aguilar as Chairman. Edward Sokolowski from ASCE,
Jack W. Pumpelly from ACSM, and William T. Pryor from ASP were members
of the Committee. William A. Radlinski, Chairman, Executive Committee,
Surveying and Mapping Division, ASCE, arranged for the joint effort
through correspondence with the Presidents of ACSM and ASP. William A.
White was contact member from the Executive Committee.

An analysis of past and present practice in uses, scales, and
contour intervals of maps for various engineering functions was made by
Aguilar with the assistance of students at Kansas State University in
1970. He found a consistent correlation of these factors. This
analysis involved a study of several hundred maps compiled for a wide
variety of engineering and associated functions. Many of the maps
studied were assembled by Pumpelly. Some of the results of this effort
were presented in the Journal of the Surveying and Mapping Division, in
July 1972 (9). These included a table showing the range of map scales
and contour intervals appropriate for basic map uses (in the article,
map uses were referred to as map types) and a means of selecting map

5

scales and contour intervals to use in various stages of performing engineering functions, considering the proposed and current land use and the character of the topography. Also included in that Journal article was a basic outline of the special publication as contemplated by the Committee at that time, and notes on the planned principal content of certain chapters.

Most of the work on the special publication was suspended after 1973 because of changes in the obligations of Task Committee members. Subjects of chapters, in preparation at that time included Map Accuracy, Map Content, Map Symbols, and Map Availability. The chapter on Map Accuracy, presenting Engineering Map Accuracy Standards (EMAS) as an alternative to the U.S. National Map Accuracy Standards (USNMAS), proposed by Merchant, had been submitted to the Task Committee. This chapter was the subject of considerable deliberation, as was a review and reassessment by Pryor of the results of Aguilar's study and their application to the selection of map scales and contour intervals.

Work on the special publication was resumed in 1977 under the leadership of Donald R. Graff, then Vice Chairman of the Executive Committee, Surveying and Mapping Division. His objective was to make the results of the previous studies and deliberation available to the profession. At his request, Merchant, Pryor, and Feldscher, all of ASCE, agreed to serve again on the Task Committee. Formal coordination of work on the special publication with ACSM and ASP was not resumed although the exchange of ideas and information with members of those organizations continued. Feldscher became Task Committee Chairman in October 1977. Robert L. Brown and Robert F. Turner joined the Committee in 1977. Brown was assigned responsibility for a chapter on Map Availability. In 1978 Robert P. Jacober agreed to serve as an advisor to the Committee and to prepare a chapter on Map Content and Symbols. John D. Bossler replaced Graff as contact member from the division Executive Committee in October 1979. Papers by Brown (1) and Jacober (4) on their chapters were published in the November 1980 Journal of the Surveying and Mapping Division as was a paper by Feldscher on the proposed purpose and scope of the special publication (3).

While the Task Committee authorization was terminated at the end of the 1979 fiscal year, work on the special publication was continued by the Committee on Cartographic Surveying. Merchant modified his presentation of the EMAS on the basis of further study and consultation and prepared the chapter titled, 'Engineering Map Accuracy Standards and Testing'. Pryor prepared a new approach to the selection of maps which he presented in two special publication chapters titled, 'Maps for Engineering and Associated Work by Stages' and 'Selection of Maps for Engineering and Associated Work'. After review and editing by the Committee, the chapters prepared by Merchant and Pryor were submitted in July 1981 to the Executive Committee, Surveying and Mapping Division, for review.

In a meeting on November 6, 1981, the Committee reviewed and discussed the comments and critiques on the chapters prepared by B. Austin Barry, Gunther H. Greulich, G. Warren Marks, and Kenneth Hunter for the Executive Committee. The following actions agreed on at the

6

meeting were taken: Pryor and Merchant modified the content of their chapters, Feldscher reviewed and edited all chapters, Bossler directed preparation of a final draft of the special publication and Morris M. Thompson performed a technical review prior to publication.

Acknowledgments and Conclusions

This special publication was prepared to assist in communication between engineers in mapping, construction and other fields. A realization of this aim, even in part, will make the effort worthwhile. The work included the development and refinement of procedures for map selection and for specifying and testing map accuracy, the research of current practice on the selection and application of map symbols, and the identification of existing map products and their sources.

The assistance of many individuals in this long-term project is acknowledged and appreciated. Those whose efforts are noted in the preceding Background statement include: Aguilar, Barry, Bossler, Brown, Feldscher, Graff, Greulich, Hostrup, Hunter, Jacober, Marks, Merchant, Pryor, Pumpelly, Radlinski, Sokolowski, Thompson, Turner, and White. Others who contributed to the preparation of the manual include: Richard C. Auth, Ralph M. Berry, Fred Brownworth, G. Martin Burdette, Dennis G. Lair, William J. Monteith, Michael Pavlides, George Reed, and Robert P. Vreeland.

The publication of this special publication, a project of the Committee on Cartographic Surveying, Surveying and Mapping Division, ASCE, provides a source of information about maps and mapping which should be of value to engineers. Updating of the contained information in future issues of the Journal of Surveying and Engineering and in new special publication editions is planned by the Committee.

The Committee owes special thanks to Mrs. Bonnie S. Maynard. Her dedicated efforts in supervising and participating in the typing, proofing, and assembling of the camera-ready copy for the special publication were essential to its publication.

APPENDIX. - REFERENCES

1. Brown, R. L., "Proposed Manual on Selection of Map Uses, Scales, and Accuracies for Engineering and Associated Purposes: Map Availability - Chapter VI," Journal of the Surveying and Mapping Division, ASCE, Vol. 106, No. SU1, Proc. Paper 15856, November, 1980, pp. 149-177.

2. Definitions of Surveying and Associated Terms, Manual No. 34, by a Joint Committee of the American Society of Civil Engineers and the American Congress on Surveying and Mapping, ASCE, 1978, pp. 100-103.

3. Feldscher, C. B., "A New Manual on Map Uses, Scales, and Accuracies," Journal of the Surveying and Mapping Division, Vol. 106, No. SU1, Proc. Paper 15853, November, 1980, pp. 143-148.

4. Jacober, R. P., Jr., "Map Content and Symbols: Chapter V of Proposed Manual on Map Uses, Scales, and Accuracies for Engineering and Associated Purposes," Journal of the Surveying and Mapping Division, ASCE, Vol. 106, No. SU1, Proc. Paper 15851, November, 1980, pp.41-72.

5. Jacober, R. P., Jr., "Standard for Symbology on Engineering Scale Maps," Journal of the Surveying and Mapping Division, ASCE, Vol. 107, No.SU1, Proc. Paper, 16651, November, 1981, pp. 21-24.

6. Official Register, ASCE, 1982, pp.151.

7. Pryor, W. T., "Surveying and Mapping, and Environmental Engineering," presented at the February 25-March 1, 1963, ASCE Environmental Engineering Conference, held at Atlanta, GA.

8. Pryor, W. T., "Specifications for Measuring and Mapping Photogrammetrically, "presented at the May 31-June 11, 1971, Eleventh Annual Photogrammetry Short Course at the University of Illinois, held at Urbana, IL.

9. "Selection of Maps for Engineering and Planning," by the Task Committee for the Preparation of a Manual on the Selection of Map Types, Scales, and Accuracies for Engineering and Planning of the Committee on Cartographic Surveying, Surveying and Mapping Division, ASCE, Journal of the Surveying and Mapping Division, ASCE, Vol. 98, No. SU1, Proc. Paper 9073, July, 1972, pp. 107-117.

CHAPTER 2. MAPS FOR ENGINEERING AND ASSOCIATED WORK BY STAGES

By: William T. Pryor,[1] Fellow, ASCE

INTRODUCTION

The purpose of Chapter 2 is to present the basic background by successive stages of the representative engineering and associated work for which maps are essential. The characteristics of maps which may be selected from available sources or must be designed and compiled for a specific use are discussed and related to the different stages.

There is no map that will fulfill every need. This is so because each map is limited in its scale, in its scope, and in the character and density of detail which can be delineated adequately throughout its area of coverage. Moreover, map needs are not the same from one engineering project or associated undertaking to another. Thus, all aspects of the work to be done must be known and considered before the requisite engineering and associated work by stages is undertaken.

Basic data pertaining to the contour interval and scale of topographic maps and to the scale of other maps, criteria for their design and compilation, or selection, and procedural principles applicable thereto are provided in Chapter 3.

The broad uses of maps essential for accomplishing engineering and associated work by stages are presented herein, and pertain to these five major fields of activity for each of which a general description follows:

Cadastral Operations
Geodetic Surveys
Hydraulic and Hydrologic Development
Land Development
Transportation

1. The cadastral operations include cadastral surveys, land acquisition, land evaluation, land subdivision, and tax assessment activities.

2. Geodetic surveys comprise placement of adequate station markers for basic horizontal and basic vertical control, and placement of station markers and/or the selection of suitable features to serve as horizontal and vertical supplemental control, and all surveying to an accuracy appropriate for computing the three-dimensional position on the surface of the earth of each of such station markers and/or features.

3. Hydraulic and hydrologic development pertains to drainage and reclamation projects, sewer and water distribution systems, dam,

---

[1] Engineering and Surveys Consultant, Arlington, VA 22205

reservoir, canal and hydroelectric facilities, and conservation and protection of wet lands.

4. Land development comprises industrial, residential, recreation, agriculture, and associated projects; and extends to urban zoning, subdivision development and control, land use, and redevelopment projects.

5. Transportation includes roads, highways, and streets; railroads and mass transportation systems; tunnels; surface and underground transmission lines; pipe lines; airports—their runways, appurtenances, and facilities; and waterfront facilities.

There are significant uses or needs for maps by stages within each major field of activity which require special and separate consideration—usually beginning with planning and ending in construction, installation and use, or other end result, as applicable.

Each field of activity requires identification and use of whatever maps are needed and beneficial within one or more of the series of coordinated engineering and associated work accomplishment stages, which are:

1. Planning;

2. Survey of Area—a general reconnaissance to determine alternatives;

3. Survey of Alternatives—a detailed reconnaissance to select one alternative;

4. Survey for Design—an accurate, detailed survey of the selected alternative;

5. Location Survey and Construction Survey—an actual staking on the ground of whatever had been designed for acquisition of rights of way [each needed parcel of land], and for construction, use, and maintenance, or improvement and/or reconstruction; and

6. As-built survey wherever necessary upon completion of construction and before backfilling trenches and so forth.

Some projects within the major fields of activity require maps throughout each successive stage, whereas others may require maps for only one stage. Presentation of the underlying principles for selection and use of maps in each stage is the objective of this chapter.

TYPES OF MAPS

Each type of map has a name identity which describes its characteristics. The four types of maps are: (1) Sketch, (2) Photographic, (3) Planimetric, and (4) Topographic. Each map requires appropriate symbolizations and labelings to categorize and identify significant details, such as soil classes and their boundaries,

10

principal classifications of land use, configurations of topographic features, drainage, and vegetation.

Use of the word 'photographic' as the descriptive identity of the second of the four types of maps includes the several commonly named maps which have the photograph as their significant base. These are mosaics of the uncontrolled, semicontrolled, and controlled classes, the orthophotograph, and the orthophotomap. Since the 1920's, it has been possible to combine some aspect of the photographic with the other three types of maps for amplification and better presentation of many details. Within recent years the orthophotograph and orthophotomap are used most, when contours are added, a topographic map results, in which their photographic images substitute for planimetry. In other cases, photographic details are considered 'cluttering' in effect and are not desirable. The choice is governed by requirements.

Although each type of map is designed and compiled for a specific use, topographic maps, which contain both planimetry and contours, are used most widely for nearly all engineering, planning, and associated work on a stage - by-stage basis. The photographic detail is added to topographic maps, as well as to the sketch and planimetric maps, when and wherever advantageous.

CONTOUR INTERVAL AND MAP SCALE

Contour interval and map scale are the principal parameters considered in selecting and/or compiling topographic maps; whereas, scale alone generally governs in the selection from available sources or compilation of the other types of maps. The type or character and intensity of land use, associated with the type of topography and vegetation, combine to affect both the contour interval and scale of topographic maps compiled or selected from a suitable and available source for use within each of the foregoing major fields of activity.

Many years of experience in the design, compilation, and use of maps, and the examination of a large number of maps throughout their range in contour interval, scale, and use within each of the five major fields of activity have led to the determination of four different criteria which have significant influence on maps. Together they are governing criteria. Collectively or separately they dominate in the selection of, or in the surveying for and compilation of, each map for its specific use, depending upon the stage of use and type of facility or works for which the engineering and/or planning is required. The criteria are:

1. Class and degree, as well as the extent, of land use development--existing and anticipated or planned;

2. Type of topography: Flat, gently rolling, rolling, hilly, mountainous;

3. Type, height, and density of the vegetation; and

11

4. Stage of use of map within each of the identified fields of use.

STAGES IN THE USE OF MAPS

To accomplish project work within each field of activity requires engineering throughout its planning, the initial stages of surveying, design, location surveying, construction, use, and maintenance. First, need is established. To do this, data pertinent to making decisions must be acquired and used effectively. Nothing should be overlooked. Once need has been verified and evaluated, all subsequent work is accomplished in the sequentially coordinated stages previously identified. That which follows comprises a brief outline of the necessary engineering work and the normal sequence within each stage in which such work is undertaken with examples in the transportation field.

1. PLANNING

Planning is the stage wherein all needs are ascertained and evaluated, and a decision is made to proceed or not to proceed, considering costs and other relevant factors. An example within the field of transportation for this stage pertains to highways, their bridges and other structures. The general principles applied within this example are sequentially applicable within each of the other four major fields of activity identified in the introduction to this chapter.

Surveys are made and relevant data therefrom are used to determine the present traffic and the anticipated traffic of 20 years hence. Maps are used to portray graphically the details of all traffic.

The portrayal of traffic is done for the number of vehicles within a specific period of time, as an hour, a number of hours, the highest hour, a day, and so forth, and the classes of traffic—passenger, bus, and truck of various types and weights—both existing and anticipated, between travel points, be they close to each other or far apart. This procedure is applicable to determinations for ascertaining the need for a new highway as well as the necessity for improving an existing highway to increase its capacity or to reduce its traffic congestion and hazards.

Once the termini of the new highway to be constructed, or the segments of an existing highway to be improved, are known, and the class of highway required has been determined, planning is ended.

The class of highway defined in the planning process is the one which will provide travel in safety, comfort, and convenience, whether the highway has full, limited, or no control of access.

The actual class of highway as determined in the planning stage for its location, design, and construction and/or improvement, is defined by the lane width and number of lanes, horizontal curvature, gradients, vertical and horizontal sight distances, and so forth, as required for use by traffic at its design speed.

12

All subsequent engineering stages pertain to determining where the highway should be constructed on the ground for most effective use by the traffic traveling between its termini and intermediate access points.

Within planning, there are five principal categories of use of topographic and other maps. These are sequentially based upon the magnitude of the area encompassed and the character and number of details which must be available for consideration in the planning process. Proceeding with maps for planning in which the contour interval is the largest and the scale is the smallest, and ending with maps in which the contour interval is the smallest and the scale is the largest, the name of each planning category is sequentially: International, National, Regional, Local, and Micro. In Chapter 3, the five will be explicitly defined; also data, principles, and procedure for selection of the map to be used in each, especially for the last three, will be featured therein by reverse order.

2. SURVEY OF AREA.

Survey of Area is the stage in which the initial engineering work is undertaken to determine all feasible alternatives of location for the facility. Also a report is prepared outlining and appropriately explaining the alternatives by route or site, functional advantages and disadvantages, and anticipated costs of right-of-way, engineering, construction and/or improvement, or other costs which may be incurred for the facility. This is an essential stage, which should be completed to ascertain and make known each feasible alternative for subsequently selecting the one which will have the least cost of construction and cost of use and will best fulfill all use requirements on a costs-benefit-ratio basis.

The survey of area relies extensively on steroscopic examination and interpretation and photogrammetric use of vertical aerial photography coverage of the entire area of concern, augmented by use of appropriate maps, to obtain essential qualitative information and quantitative data for evaluating all feasible alternatives within that area for the proposed facility. For example, canals, transmission lines, pipe lines, railroads, and highways require route corridors of adequate width and appropriate changes in direction to fit within the topography and land use from one terminal point to another.

Among the considerations and controls determined in this stage [for which essential coverage by aerial vertical photographs and maps of adequate scope, scale, and accuracy must be obtained and used (augmented where and when advantageous or necessary by use of other photographs and on-the-ground inspections)] are:

1. The environment, existing and as it will become after the facility has been sited or the route chosen, surveyed, designed, and constructed;

2. Land uses within the area of concern, both existing and potential, and identification and classification of each from the

13

viewpoint of being affected beneficially or adversely;

3. The topography, and its drainage and vegetation—whether existing or needed;

4. Classes or types of soil on both a general area and specific site selection basis, and identification of areas within the alternative corridors or sites where soil is so unsuitable that such areas must be avoided;

5. Places subject to excessive erosion or flooding; and

6. All other factors having an influence upon location of each feasible site or route, including: Faults and earthquake zones; existing mines and shafts; wells; utilities, both above ground and underground; and so forth.

The word controls precedingly identified has multiple meanings. First, it may be negative or positive. Whenever it is negative it means the areas of finite sites which should be avoided within the identifiable land use, topography, and drainage. Examples are a cemetery; a public building, as the Capitol; a land slide; a swamp; severance of valuable farming lands; and an identifiable snow drifting or water flooding area. Whenever it is positive it is applicable to areas or finite sites which should be utilized or served. Positive controls include: Stable slopes within mountainous regions; the lowest and most directionally positioned mountain pass; the best directional positioning of a bridge site for the highway over a river, another highway, railroad, canal, and the like; the place where an interchange can be best fitted to the topography to connect with streets or other highways. Next, controls mean the objective of fitting the proposed facility within the land use, the topography, cross drainage and classified soil areas; and environment, vegetation, and weather created advantages where it will fulfill its intended purpose with best service for least cost of construction and use. Innumerable examples might be cited. Perhaps the few mentioned herein will suffice to give an insight into the real meaning of controls.

Continuing with the highway project example, the engineering and associated work done in the survey of area stage will usually include the following procedures and activities.

All feasible route alternatives between the designated terminal points, be they close together or far apart, will be determined to ascertain where, within the topography and land use throughout the area, corridors exist which are feasible for the proposed highway. Such engineering work will require thorough examination of the broad area—four-tenths to six-tenths as wide as the area is long between the termini—by use of appropriate maps and augmenting qualitative information and quantitative data from aerial photographs. Investigations on the ground at all critical places, with the maps and photographs in hand, are also advisable. Each route alternative is determined on the basis of the class of highway to be located and/or improved, and comprises a corridor which is sufficiently wide and circuitous within the bounds of all topographic and land use controls for the proposed highway. Moreover, each alternative is ascertained as if it were to be the one along which the highway will be constructed after the design has been completed and detailed construction plans are prepared.

In doing such work, each one of these significant aspects will have its governing influence throughout the process of determining the location of each route alternative:

1. Curvature and gradients

2. Width and number of traffic lanes, and median if any.

3. Limitations in fulfilling the foregoing for the class of highway by:

   (a) Land use, topography, soil, drainage, vegetation, and seasonal weather; and

   (b) The environment, existing, potential, and/or anticipated.

4. Probable cost for the highway on each route alternative for:

   (a) Rights-of-way;

   (b) Engineering and construction;

   (c) Maintenance; and

   (d) Vehicle operation—the total for all traffic.

A report is prepared comprising a complete presentation of all route corridor alternatives, together with pertinent facts about each on the basis of the foregoing. In all instances, of course, the maps of suitable scale, scope of coverage, and accuracy that had been used in accomplishing the engineering work and doing the reporting will be identified, as well as the photography used.

Inasmuch as the Survey of Area encompasses both large and small areas, according to the character and magnitude of the undertaking, this stage is often thought of as being a part of planning—either National or Regional, and in other instances it is sometimes thought of as Local Planning. Certainly, the contour interval and scale are important parameters in selection of maps for use in accomplishing the engineering and associated work in each category of planning, as well as in the survey of area for determination of feasible alternatives. Whenever essential maps for accomplishing the planning and survey of area are not obtainable from a suitable source, it may be necessary to accomplish the topographic surveys on the ground or use appropriate photogrammetric methods for compiling the maps.

15

3.   SURVEY OF ALTERNATIVES.

In the Survey of Alternatives stage, all feasible routes or sites that were ascertained and located in the preceding stage are compared. The comparisons are done in adequate scope and sufficient detail for selecting one from among the many which will fulfill all considerations and best serve within the scope and purpose for which the proposed facility is needed, and do so for the least costs of its engineering, right-of-way, construction, use, and maintenance. Another systematization procedure for comparing the alternatives in this engineering stage is to think of the work as Local Planning, if the size of the undertaking is large; and Micro Planning, if the alternatives are small in size and if their details are complicated and intricate.

Using again the example of a highway engineering project, procedures subsequently presented would be followed. In this survey stage, pertinent quantitative data and qualitative information are obtained by use of the selected, or compiled maps of each route, augmented by use of aerial and other photography.

The feasible route alternatives determined in the preceding stage are compared. As previously mentioned, the comparisons are made in adequate scope and sufficient detail to select the best from among the many. The best route, of course, is the one along which the required highway can be surveyed, designed, and constructed where it will appropriately serve the planned-for traffic at the design speed, and do so for the least costs of construction, maintenance, and use.

In accomplishing all of such engineering work, each of the following are verified and evaluated:

1. The significant controls of: Land use, topography, drainage; the soil areas which can be used and which should be avoided; the environment, both existing and potential, anticipated; seasonal weather, and exposure associated with the bordering vegetation and topography.

2. The possibilities of fitting the highway during design and construction, and doing it feasibly and economically, along each feasible route to:

   (a) The existing and potential use;

   (b) The topography, with its drainage, and soil and/or rock; and

   (c) Traffic service requirements, terminus to terminus, and between all other places of access.

3. Costs, to a reconnaissance degree of accuracy, of a highway of the class required along each of the feasible route alternatives, for

(a) Engineering;

(b) Rights-of-way;

(c) Construction;

(d) Maintenance; and

(e) Vehicle operation.

4. Local response at Public Hearings regarding each of the route alternatives.

5. Recommendations, with evaluations, and designation of the selected route.

6. Timetable for accomplishing subsequent:

(a) Engineering, including: (1) surveying before, (2) design, and (3) surveying afterward;

(b) Right-of-way procurement; and

(c) Construction, with its associated surveying: (1) before, (2) during, and (3) afterward.

7. Or, if found necessary or desirable, termination of further work.

The Survey of Alternatives is a transition stage. It is the stage in which the many feasible alternatives are compared, and the best is ascertained from the viewpoints of achieving the least damage and interference to whatever exists within its bounds and along its borders, and attaining the best service from the facility after it has been completed and is in use. Throughout this stage the contour intervals and map scales come within those applicable to Local and Micro Planning, and can range into the smaller contour intervals and larger scales used in General Design.

4.   SURVEY FOR DESIGN.

This is the stage in which a preliminary survey is made of the selected route or site to obtain finite qualitative information and quantitative data. Seldom, if ever, will existing maps and aerial photography suffice. Consequently, this survey must be made.

All surveying is done in the detail, of the scope, and to the accuracy necessary for obtaining the qualitative information and quantitative data needed to do the engineering design and prepare plans in the detail and of the scope necessary for constructing the proposed facility. The plans comprise a complete delineation of all design details on the maps compiled and used for both general and critical design. General design pertains to overall details of construction throughout the project. Critical design is confined to specific sites

17

where topographic and land use problems are acute, and the design details must be presented at much larger scale than is possible from general design. The surveys for compiling these required maps can be done on the ground or the compiling can be done by photogrammetric methods if the ground is not obscured too much by tall and dense vegetation. Choice of method is usually made on the basis of whichever will be the most efficient and effective for the types of land use, topography, geologic structure of the ground, soils, and vegetation encountered, associated with the scope, details, and accuracy required throughout each needed map.

Regardless of which surveying method is employed, the end result will be topographic maps, or planimetric maps supplemented by essential vertical dimensional data in the form of cross sections and/or spot elevations. The selection of contour interval by scale for topographic maps, and the selection of scale of all other maps used with profile, cross sections and spot elevations for design are presented in Chapter 3.

Each map used for design should be augmented by stereoscopic examination and interpretation of aerial photographs of adequate scale—in the representative fraction scale range of 1:12,500 to 1:2,500, or the range of 1,000 feet per inch to 200 feet per inch. Occasionally, the scale of the map-augmenting photography may be as small as 1:25,000, or 2,000 feet per inch throughout large mountainous and other types of undeveloped areas.

It is not the purpose of this chapter to present all aspects of such surveys, whether by on-the-ground measuring of angles and distances (both horizontal and vertical), by aerial photogrammetric methods, or by a combination of both methods. Regardless of method employed, the resulting topographic maps and/or augmenting planimetric maps with profile, cross sections, and spot elevations must be adequate in scope and have the scale, details, and accuracies essential for the design and the resultant construction plans. As mentioned initially, nothing should be overlooked, omitted, or ignored.

5.   THE LOCATION SURVEY AND CONSTRUCTION SURVEYS

All work throughout the foregoing stages is focused toward attaining the end result—the constructed facility or other objective. But, after the general design and augmenting critical design have been completed and the detailed construction plans have been prepared, the facility cannot be constructed until it has been staked on the ground.

There are two different ways in which surveys for that purpose are identified. One is called The Location Survey and the other The Construction Survey. In some instances both terms are combined to mean Location and Construction Survey. In others, The Location Survey precedes the Construction Survey.

Regardless of which survey name is used, the work comprises the surveying which must be done accurately to set in the ground the right-of-way and construction stakes to define for construction the details and their limits for each aspect of the facility.

As an example, within highway engineering, The Location Survey comprises staking the centerline of the designed highway on the ground and measuring its profile; and the Construction Survey pertains to measuring cross sections and setting slope stakes, replacing the centerline as necessary during the construction operations, and setting grading stakes as the work progresses to serve as the finite limits for finish grading, paving, and so forth.

The rights-of-way boundaries may be staked as part of The Location Survey or staked later, either during or after construction. The sequence is optional, according to which procedure is most efficient and effective under the topographic and land use conditions existing and the types and densities of vegetation throughout the limits of the facility.

The contour intervals and map scales, and augmenting information, used in this stage are the same as those employed in the Survey for Design stage. This is because the detailed construction plans comprise working copies of the maps used during design, and contain all details of the design superimposed upon the maps encompassing the selected site and/or route of the facility.

6.   AS-BUILT SURVEYS.

Each survey of the engineering works, after all construction is completed, comprises essential angle and distance measuring, both vertical and horizontal, for defining the actual location (X, Y, and Z position) of the completed facility or works, with its appurtenances.

CHAPTER 3. SELECTION OF MAPS FOR ENGINEERING AND ASSOCIATED WORK

By: William T. Pryor,[1] Fellow, ASCE

INTRODUCTION

As demonstrated in Chapter 2, the use of maps is essential throughout all stages of engineering and associated work. Appropriate topographic maps, or augmenting maps of the sketch, photographic, or planimetric type, accompanied by use of cross sections and/or spot elevations for dimensional definition of the configurations of the surface of the ground, must be obtained and used. They are required in each successive stage: Planning, location, design, construction, use, and maintenance of each facility or works.

This chapter speaks of contour interval, scale, and content of topographic maps; and scale and content of other maps. Also presented are principles, project criteria, and procedures to assist engineers in selecting basic parameters for designing and compiling maps, and for selecting maps from appropriately available sources. Thereby, each procedurally selected and/or designed and compiled topographic map, and all augmenting maps wherever needed, will fulfill planning and engineering requirements. Moreover, an effort is made to provide engineers who are not specialists in surveying and mapping with practical factual information and criteria pertaining to contour interval and map scale in relation to land use, topography, stage of map use, character and intensity of detail required on the maps, along with an orderly procedure for determining the contour interval and scale of map which will fulfill each planning and engineering requirement.

This is done by the use of charts having a logarithmic grid, with companion tables. In one set of charts and their companion tables, the contour interval for each topographic map is in meters, and its scale is expressed as a representative fraction. In another set of charts and tables, the contour interval for each topographic map is in feet, and its scale is in feet per inch. Parameters of correlation are provided according to engineering stages. The parameters are applicable for both current and planned land use, and for the topography throughout each area of concern.

To define and present the functionally supporting relationship between the contour interval and scale of topographic maps used throughout the various stages of planning and subsequent engineering, the experience of many map designers, compilers, and users has been incorporated. Basically, the information presented herein was derived from an examination and analysis of the contour interval, scale, and cartographic detail of maps collected from numerous sources, as well as consulting various specialists. Topographic maps were examined,

[1] Engineering and Survey Consultant, Arlington, VA 22205

classified, and analyzed within the range of contour interval and scale of maps used throughout the stages of planning and engineering by governmental, consulting, and commercial users within the United States, and from several sources in other countries. The findings and resultant presentations herein are a summary of past and present practices, supplemented by personal knowledge obtained throughout many decades of designing, surveying for and compiling, and using maps.

There are constituent functions, within each stage of engineering work, for which maps are required. The 'functional uses of maps' or 'map uses' are described and identified in the first part of this chapter. Practical limits in contour interval and map scale for each map use are shown graphically on charts 1 and 2 and listed in Table 1.

Near the end of this chapter, several examples of map selection are presented. The examples are augmented with salient facts pertaining to correlation of photogrammetric technology, instruments, and systems. The correlation pertains to those which (1) are in use, (2) can be used, and (3) may be developed for use in the future. This illustrates that each choice will bestow the capability and reliability of making each essential measurement and compiling each required map to requisite accuracy.

Before any map is used its accuracy should be known. For most maps, and especially those of large scale, the actual reliability of each map can best be ascertained by making field tests which will lead to factual knowledge of its accuracy. Details pertaining to engineering map accuracy standards and testing are presented in Chapter 4.

MAP USES

The uses of topographic and other maps are presented in order from small contour interval and large scale to large contour interval and small scale. The map uses are word defined and numerically and alphabetically arranged on the charts and in the tables by similar order. The reason for this is the ease with which such relationships can be portrayed sequentially in such an order on graphical charts and in tabular form. This is in reverse, of course, from the sequential order of map use for planning and continuation of map use throughout the successive stages of engineering, which begin with large contour intervals and small scales for the topographic maps, and progressively proceed to small contour intervals and large scales.

[We have seen in Chapter 2 that, after planning, the sequential engineering stages comprise survey of area to determine feasible site or route alternatives, survey of the alternatives to select one from the many, survey of the selected alternative for compiling adequate maps to design the facility and thereby prepare detailed construction plans, location survey and construction surveys, and ultimately the as-built survey.]

DESCRIPTIONS OF THE MAP USES

The descriptions of map uses herein are by the sequential engineering stages. The purpose is to provide background information for correlation with the subsequently presented principles and procedures of selecting maps by contour interval or cross sections and/or spot elevations, by character and intensity of cartographically delineated planimetric details, and by scale.

In subsequently presented Charts 1 and 2, the extremes in contour interval and scale for all topographic maps are exemplified, and are numerically listed in Table 1. First, the extremes will be presented for each stage, with only general rounding, as applied within Table 1. Next, the principles and procedures of selecting the contour interval and scale for each map required in the successive stages will be featured, along with a factual presentation of the feasible and practical rounding of contour interval and scale.

MAPS FOR DESIGN.

Topographic maps required for design are always compiled and used with a contour interval small enough and a scale sufficiently large to portray the configurations of the surface of the ground, as well as to contain all essential planimetry in adequate detail and accuracy. This is necessary for accomplishing the design and preparing the detailed construction plans. Each map must also have adequate scale for superimposing on it all design details necessary to portray the end result that is to be staked on the ground for construction.

For design of some facilities, however, planimetric maps are used instead of topographic maps. In such instances all needed vertical dimensions are provided in the form of appropriately spaced and sufficiently accurate cross sections and/or spot elevations. An example of such practice is the design and preparation of detailed construction plans for a highway along the route selected from feasible route alternatives across a large level area like the beds of the extinct lakes of Nevada and Utah, and the extensive Rio Atrato Swamp in northern Colombia. Contour lines in such flat areas are so far apart on the map that they, in effect, become nearly meaningless.

Dimensional data provided by all maps must be augmented by essential qualitative information cartographically symbolized and delineated on the maps; this information is supplemented in the design office by the use of aerial and other photographs and by on site inspections with the maps and photographs in hand.

A. Maps for Critical Design.

If the project or facility for which the design must be made has restrictions in area, orientation, position, and/or elevation, it has the critical design classification. A typical critical design project is a traffic interchange within a highly developed industrial or urban area. Another example is the surface and underground drainage for a housing or industrial complex. A third is the design of a large

22

structure, such as a bridge; others are buildings for an industrial complex, or an urban or suburban housing development.

The topographic maps most suitable for critical design have a contour interval range of 0.01 meter to 4 meters, and a scale range of 1:25 to 1:600; or a contour interval range of 0.03 foot to 10 feet, and a scale range of 2 feet per inch to 50 feet per inch. The choice depends on the topography, drainage, soil, land use, other major features within the area of concern, and detail required for the design and construction plans.

B. Maps for General Design.

When projects or facilities, small or large, are to be constructed or improved, general design is necessary. General design comprises determining the site for the specific project or facility within bounds of its route and/or area of survey; and appropriately delineating the details of all aspects of the project and/or facility on the maps so that it can be staked on the ground, along with its rights-of-way.

Adequate accomplishment of all essential engineering in this design stage requires topographic maps providing vertical dimensions from contours of small enough interval or planimetric maps with cross sections and/or spot elevations appropriately spaced. The maps must also contain planimetrically each horizontal configuration, position, and dimension essential for portraying representatively the ground surface and all other topographic and land use features. In addition, vital qualitative information regarding existing utilities above and below ground, property boundaries, and land ownership throughout the area of concern at an effective scale will be needed.

Examples of projects requiring general design are irrigation and other water distribution systems, canals, dams, water front facilities, hydroelectric facilities, urban and suburban housing, industrial complexes, mass transit systems, railroads, and highways.

For general design, the contour interval ranges from 0.06 meter to 8 meters at map scales ranging from 1:500 to 1:2,500; or the contour interval ranges from 0.2 foot to 25 feet at map scales ranging from 40 feet per inch to 200 feet per inch; all depending upon the type of topography, drainage, geology, vegetation and its height and density, land uses, and all relevant details to be depicted on the maps after the design has been completed.

MAPS FOR PLANNING.

In planning for specific engineering works or other facilities, a large range exists in the contour interval and scale of the topographic maps that are used, regardless of whether the planning is micro, local, regional, national, or international in character. The maps are used for representing all details having significant influence on decisions to be made during the planning process, and for adding, after the planning has been accomplished, every detail necessary for depicting results.

Topographic and other maps used for planning purposes contain vital
details on position, bounds, and characteristics of geology; land uses
by man and by nature, including their variations and intensities;
agriculture, its types and production; drainage, both general and
detail; networks of transportation systems, as railroads, highways,
airways, canals, electric power and telephone lines, and irrigation;
geodetic surveys of the horizontal and vertical types; cadastral
surveys, with land ownership where essential; soils by types and
boundaries; urban, suburban, and industrial complexes; and all other
essential information and data. Actually, each map should contain
whatever is required for accomplishing the planning to be done,
consistent with the purpose for which the end result is required.

Each of the various types of planning has its particular
requirements of detail, contour interval, scale, accuracy, and scope of
coverage, whether it be micro, local, regional, national, or
international in character.

Maps for national and international planning usually pertain to
very large areas. Topographic maps used for such planning have contour
intervals of from 1.5 meters to 3,000 meters at scales of from 1:125,000
to 1:60,000,000; or the topographic maps have contour intervals of from
5 feet to 10,000 feet at scales ranging from 10,000 feet per inch to
5,000,000 feet per inch [Refer to Table 1.], or scales of from 2 miles
per inch to 1,000 miles per inch. Maps for those two types of planning
will be presented only in a general manner and not in detail within this
chapter. Maps for engineering and associated work are featured herein.

C. Maps for Micro Planning.

Micro planning comprises the detailed comparison of alternatives by
site or by route so as to select one for the facility or works to be
subsequently designed and constructed. Such planning includes depiction
of details and features [by use of topographic maps--supplemented, as
necessary, by other types of maps] pertaining to existing facilities to
be altered for expansion or improvement, or to be removed and replaced;
for planning and accomplishing cadastral and geodetic surveys; for land
evaluation and assessment; and for all engineering work required to
determine where best to situate the works or facility.

Thus, micro planning is essential for making the site or route
selection from among the alternatives within critical areas as well as
large general areas for subsequent critical or general design,
respectively, which precedes staking on the ground, construction, use,
and maintenance of the facility or works.

Topographic maps used in micro planning have contour intervals of
0.1 meter to 20 meters and scales ranging from 1:1,250 to 1:12,500; or a
contour interval ranging from 0.4 foot to 60 feet and scales of 100 feet
per inch to 1,000 feet per inch. Selection of each topographic map from
available sources, or its compilation where necessary, is based upon the
purpose of the project, the type of topography, character and intensity

of land use, soils, drainage, geology, vegetation, and all other features that will affect decisions.

D. Maps for Local Planning.

Local planning consists in determining the feasible alternative sites and/or route locations for a proposed facility or engineering works.

The size of the area of investigation for accomplishing local planning is governed by the type or character of the undertaking. Some areas are small and some are large; some are somewhat square; and others are long, narrow, and circuitous, but sufficiently wide to include all identifying location controls[a] of topography, land use, soil, drainage, vegetation, and so forth.

Examples of long-route alternatives include those for electric power transmission and telephone lines, pipe lines, canals, railroads, or highways. Local planning may also include preparations for making cadastral and geodetic surveys throughout all areas of concern. Essential qualitative and quantitative data which will have an effect upon the work must be obtained, adequately symbolized or otherwise identified, and delineated on the maps; and would include utilities, both above and below ground, within all areas of concern, especially those which are urban and suburban in character.

The topographic maps which will usually fulfill requirements for local planning have contour intervals from 0.25 meter to 30 meters at scales of 1:5,000 to 1:25,000; or have contour intervals from 0.75 foot to 100 feet and scales ranging from 400 feet per inch to 2,000 feet per inch.

E. Maps for Regional Planning.

Regional planning, whether the region is large or small, is to determine regional needs, and evaluate all aspects of a facility or other work in relation to the needs, and also to ascertain whether or not further work should be done leading to fulfilling all needs. Included within such planning are determining: (1) advantages, (2) disadvantages, (3) environmental influences of what is currently there, and (4) what the preceding three will be after the intended has become reality.

Regional planning comprises the work introduced under the heading of "1. PLANNING" in Chapter 2. This planning includes considering the environmental and social aspects of the proposal; communication, transportation, the natural resources to be utilized or to be advantageously or adversely affected; and land uses which will be

---

[a] The meaning herein for controls is the same as defined in Chapter 2 within the fourth paragraph under the heading: 2. SURVEY OF AREA.

changed and benefically or adversely affected.

Maps generally used for regional planning are the national
topographic and geologic series, State-wide planning maps, county maps,
city maps, statistical regional maps, coastal and harbor navigation
charts, and other acceptable maps. If suitable maps are not obtainable
from available sources, adequate surveys have to be made and the
required maps compiled.

Topographic maps for regional planning have contour intervals from
0.4 meter to 80 meters at scales of 1:12,500 to 1:125,000, or contour
intervals of 1.25 feet to 250 feet and scales of from 1,000 feet per
inch to 10,000 feet per inch. All other types of maps for regional
planning have similar ranges in scale.

F. Maps for National Planning.

Maps for national planning are obtained from available sources or
must be compiled. Included are maps of natural resource inventories,
communication and transportation systems, agriculture, navigation on
national waters and by air, national defense [defining boundaries on
land and limits of control over bordering waters] and so forth. Such
maps may be of the sketch, planimetric, photographic, or topographic
types, whichever will fulfill needs.

The contour interval of topographic maps used for national planning
ranges from 1.5 meters to 300 meters at map scales of 1:125,000 to
1:1,500,000, or the contour interval ranges from 5 feet to 1,000 feet at
scales of 10,000 feet to 125,000 feet per inch.

G. Maps for International Planning.

International planning includes ascertaining data for and planning
and coordinating international cooperative programs; acquiring and
analyzing population data and graphically portraying its details and its
bounds; analyzing and planning communication, transportation, and
defense systems; developing and sponsoring agricultural, manufacturing,
and environmental conservation and improvement; identification,
evaluation, conservation, and development of natural resources; and
other programs of an internationally beneficial and cooperative nature.

Topographic maps used for international planning generally have
contour intervals of 6 meters to 3,000 meters at scales of 1:1,250,000
to 1:60,000,000, or contour intervals of 20 feet to 10,000 feet, and
scales of from 100,000 feet per inch to 5,000,000 feet per inch (or
scales of from 20 miles to 1,000 miles per inch). [Sometimes the small
contour interval of 6 meters is truncated downward, by what is often
called hard conversion from a contour interval of 20 feet, to 5 meters
for personal preference.] All other types of maps used in conjunction
with such planning have similar scales.

Within many countries the maps used are purchased from the shelf
supplies of an established and functioning mapping organization, whether
a governmental agency or a private firm, unless maps from such sources

will not fulfill needs in this planning stage. Then the maps must be compiled to specifications setting forth the scope, specific details, scale, and accuracies required.

CHARTS AND TABLES OF CONTOUR INTERVALS AND MAP SCALES

There are finite relationships existing between the contour intervals and scales of topographic maps which are illustrated effectively by use of logarithmic charts and numerical tabulations. The principles of selecting such maps for effective use in each engineering and planning stage are procedurally presented through use of the charts and their companion tabulations.

CHARTS 1 AND 2.

Charts 1 and 2 are logarithmic, with the abscissa the contour interval and the ordinate the map scale. For Chart 1 the contour interval is in meters and the map scale is identified by the denominator (N) of the representative fraction expressing map scale. For Chart 2 the contour interval is in feet and the map scale is in feet per inch.

Each chart contains three lines sloping upward from left to right. The first of such lines represents the minimum contour interval, the second the median contour interval, and the third the maximum contour interval for each map scale.

Three separate equations were derived empirically for computing from the scale of maps: 1. the minimum, 2. the median, and 3. the maximum contour interval. This was done using plottings of contour intervals from the numerous topographic maps obtained and examined, and from personal experience and knowledge. The empirical equations are recorded within the headings of Table 1 for computing each contour interval: (1) in meters from the 0.6 power of the numeral (N) identifying the denominator of the map scale expressed as a representative fraction, and (2) in feet from the 0.6 power of the map scale (MS) in feet per inch. Each applicable constant, called a 'K' factor, is included for computing the extremes in contour interval, as well as for the median contour interval.

Each equation, with its applicable 'K' factor, was used to compute the data recorded in Table 1 without fully rounding or truncating the contour intervals. Also, each equation is repeated for Table 5, together with a 'K' factor for the minimum, median, and maximum contour interval in both meters and feet for each type of topography.

Both charts contain horizontal lines which define the minimum and maximum scale limits for each of the two functional categories of design, A and B, and for each of the first four functional categories of planning, C, D, E, and F. There is an overlap within each of such bounds except for category F. As an example, the overlapping demarcation of limits for the two different categories of design by map scale are 1:500 and 1:600, or 40 feet per inch and 50 feet per inch, comprising the minimum for B and the maximum for A on charts 1 and 2, respectively.

27

Use of each chart enables engineers and planners to visualize
readily the limit of contour interval and map scale for each planning
and engineering use, regardless of whether the measurement units are
meters or feet. The median indicates generally the contour interval
frequently used for each map scale within bounds designating the
sequential stages.

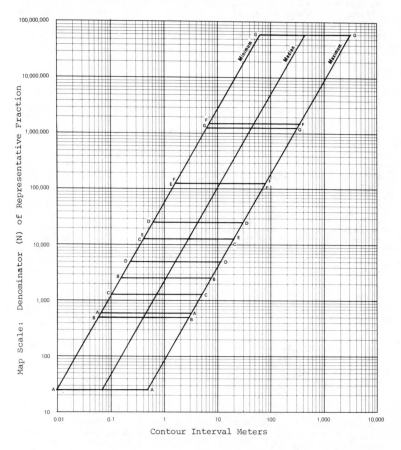

Chart 1.   Contour Interval & Map Scale Limits in Functional Use of
Topographic Maps

(Contour Interval is in Meters, and Scale is the Representative Fraction.)

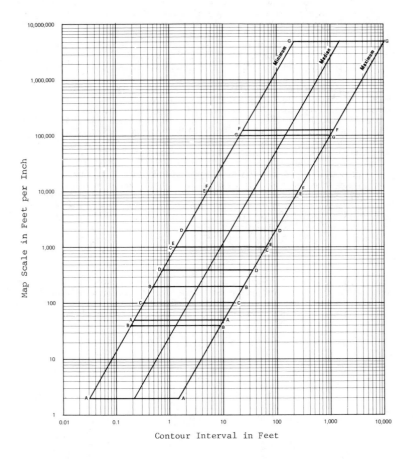

Chart 2.   Contour Interval & Map Scale Limits in Functional Use of
Topographic Maps

(Contour Interval is in Feet, and Scale is in Feet per Inch.)

29

TABLE 1.

Table 1 shows contour intervals in METERS and FEET by scale for topographic maps throughout the seven functional stages, A through G, in which the maps are used. For topographic maps on which the contour interval is in meters, the scale is indicated by the number (N) comprising the denominator of the representative fraction expressing map scale; and for such maps on which the contour interval is in feet, the scale (MS) throughout the table is in feet per inch.

To maintain computational continuity throughout Table 1, most tabulated contour intervals have a decimal fraction ending. Fractional endings would be used for contours on topographic maps: (1) if the interval must be small enough to represent all configurations of the topography at large map scales, and (2) if necessary to maintain rounded index contours. In any case, however, appropriate rounding and/or truncating would be employed to define the contour interval.

In actuality, Table 1 is a numerical listing for functional use throughout the engineering stages of topographic maps by contour intervals and associated map scale, as graphically presented on Charts 1 and 2. Each contour interval is computed for each map scale using the appropriate empirical equation with its applicable 'K' factor for minimum, median, and maximum interval in meters and in feet, respectively. For all maps not containing contours, only the map scale listings apply throughout the functional uses of maps by engineering stages.

Table 1 and the charts augment each other and are the foundation upon which all subsequent presentations and examples in this chapter are based. Table 1 is also a correlation summary in reverse order of the general stage-by-stage procedural principles presented in Chapter 2.

TABLE 1.   Contour Interval in METERS & FEET by Scale of Topographic Maps for Engineering Stages.

Scale Number (N) of Represent. Fraction 1 : (N)	Contour Interval $CI = K (N)^{0.6} = M.$			THE ENGINEERING STAGES and Functional Uses of TOPOGRAPHIC and OTHER MAPS	Map Scale (MS) Feet per Inch	Contour Interval $CI = K (MS)^{0.6} = Ft.$		
	K = .0014 Min.	K = .0098 Med.	K = .0686 Max.			K = .0204 Min.	K = .1428 Med.	K = .9996 Max.
(1)	(2)	(3)	(4)	(5)	(6)	(7)	(8)	(9)
				SURVEY for DESIGN:				
25	0.01	0.07	0.47	A. Critical	2	0.03	0.22	1.52
50	0.01	0.10	0.72		4	0.05	0.33	2.30
60	0.02	0.11	0.80		5	0.05	0.38	2.63
100	0.02	0.16	1.09		8	0.07	0.50	3.48
125	0.03	0.18	1.24		10	0.08	0.57	3.98
200	0.03	0.24	1.65		16	0.11	0.75	5.28
250	0.04	0.27	1.88		20	0.12	0.86	6.03
400	0.05	0.36	2.50		30	0.16	1.10	7.69
500	0.06	0.41	2.86		40	0.19	1.31	9.14
600	0.07	0.46	3.19		50	0.21	1.49	10.5
500	0.06	0.41	2.86	B. General	40	0.19	1.31	9.14
600	0.07	0.46	3.19		50	0.21	1.49	10.5
800	0.08	0.54	3.79		60	0.24	1.67	11.7
1,000	0.09	0.62	4.33		80	0.28	1.98	13.9
1,250	0.10	0.71	4.95		100	0.32	2.26	15.8
2,000	0.13	0.94	6.56		160	0.43	3.00	21.0
2,500	0.15	1.07	7.50		200	0.49	3.43	24.0

TABLE 1.--Continued

				SURVEY OF ALTERNATIVES:				
1,250	0.10	0.71	4.95	C. Micro Planning	100	0.32	2.26	15.8
2,000	0.13	0.94	6.56		160	0.43	3.00	21.0
2,500	0.15	1.07	7.50		200	0.49	3.43	24.0
5,000	0.23	1.62	11.4		400	0.74	5.20	36.4
6,000	0.26	1.81	12.7		500	0.85	5.94	41.6
8,000	0.31	2.15	15.1		600	0.95	6.63	46.4
10,000	0.35	2.46	17.2		800	1.13	7.88	55.2
12,500	0.40	2.81	19.7		1,000	1.29	9.01	63.1
5,000	0.23	1.62	11.4	D. Local Planning	400	0.74	5.20	36.4
6,000	0.26	1.81	12.7		500	0.85	5.94	41.6
8,000	0.31	2.15	15.1		600	0.95	6.63	46.4
10,000	0.35	2.46	17.2		800	1.1	7.9	55.2
12,500	0.40	2.81	19.7		1,000	1.3	9.0	63.1
20,000	0.53	3.73	26.1		1,600	1.7	11.9	83.6
25,000	0.61	4.27	29.9		2,000	2.0	13.7	95.6

				SURVEY OF AREA				
12,500	0.40	2.81	19.7	E. Regional Planning	1,000	1.3	9.0	63.1
20,000	0.53	3.73	26.1		1,600	1.7	11.9	83.6
25,000	0.61	4.27	29.9		2,000	2.0	13.7	95.6
30,000	0.68	4.76	33.3		2,500	2.2	15.6	109
40,000	0.81	5.66	39.6		3,000	2.5	17.4	122
50,000	0.92	6.47	45.3		4,000	3.0	20.7	145
60,000	1.03	7.21	50.5		5,000	3.4	23.7	166
80,000	1.22	8.57	60.0		6,000	3.8	26.4	185
100,000	1.40	9.80	68.6		8,000	4.5	31.4	220
125,000	1.60	11.2	78.4		10,000	5.1	35.9	251
125,000	1.60	11.2	78.4	F. National Planning	10,000	5.1	35.9	251
200,000	2.12	14.9	104		16,000	6.8	47.6	333
250,000	2.43	17.0	119		20,000	7.8	54.4	381
300,000	2.71	18.9	133		25,000	8.9	62.2	435
400,000	3.22	22.5	158		30,000	9.9	69.3	485
500,000	3.68	25.7	180		40,000	11.8	82.4	577
600,000	4.10	28.7	201		50,000	13.5	94.2	659
1,000,000	5.57	39.0	273		80,000	17.8	125	874
1,250,000	6.37	44.6	312		100,000	20.4	143	1,000
1,500,000	7.11	49.8	348		125,000	23.3	163	1,143
1,250,000	6.37	44.6	312	G. International Planning	100,000	20.4	143	1,000
1,500,000	7.11	49.8	348		125,000	23.3	163	1,143
2,500,000	9.66	67.6	473		200,000	30.9	216	1,515
5,000,000	14.6	102	717		400,000	46.9	328	2,296
10,000,000	22.2	155	1,087		800,000	71.0	497	3,481
12,500,000	25.4	178	1,243		1,000,000	81.2	568	3,979
50,000,000	58.3	408	2,856		4,000,000	187	1,306	9,142
60,000,000	65.1	455	3,186		5,000,000	213	1,493	10,452

SELECTION OF CONTOUR INTERVAL AND SCALE FOR TOPOGRAPHIC MAPS

Selection has two different meanings. The first is selection of maps from available sources. The second is selection of contour interval and scale for compilation of maps from surveys made on the ground or by photogrammetric methods and their associated control surveys, whichever is most suitable and economical, depending upon the character and condition of the topography and drainage, ground cover by vegetation, land use, and so forth throughout the area or site of survey, and the accuracy required.

Selection of a topographic map having proper contour interval and scale relationships comprises the basis for successful accomplishment. The underlying aspects of selection are given for each engineering and planning stage and functional use of topographic maps. All that follows leads to application by examples of the underlying principles, facts, and relationships. The additional tables and charts which are provided are for making procedural use of the fundamentals depicted by Table 1, from Survey for Critical Design through Survey of Area and Regional Planning.

TABLE 2.

By use of fifteen assigned identifying map group numbers, Table 2 serves as the correlator between contour interval, map scale, and the various fields of activity in which maps are used for accomplishing engineering and planning. The correlation is achieved by employing the previously defined designating letters of A, B, C, D, and E for map use to serve as the heading identity of the third to seventh column. Then a range of identifying map group numbers for the major fields of activity and types of projects (works or facilities) named within the second column under the subheading of Project were assigned within each column having the heading of A to E, respectively.

Table 2 is not complete by itself. It requires use of augmenting tables, which are subsequently identified and described. It is applicable regardless of whether the measurements made and maps to be employed are based upon the metric or foot system.

The map group numbers within each assignment in Table 2 actually comprise a range in designation for the compatible contour intervals and map scales. The map group numbers are jointly used for correlation of contour interval and map scale. This is done in column 1 for Tables 3-M and 3-T, and in column 2 for Tables 4-M and 4-T, respectively.

There is an overlap of the assigned map group number between successive A to E map use designations. This is because correlation cannot be relegated for both the contour interval and map scale to a single finite segregational boundary limit between each of the sequential stages of engineering and associated work.

TABLE 2. Map Group Numbers by Fields of Activity for Map Use Selection by Contour Interval and Map Scale.

Fields of Activity Classification	Project	Map Group Numbers for Map Use Selection A	B	C	D	E
Cadastral Operations	Boundary Surveys	4 to 6	5 to 9	8 to 11	10 to 13	12 to 14
	Land Acquisition[a]	4 to 6	5 to 9	8 to 12	10 to 13	12 to 14
	Land Evaluation	--	--	8 to 12	--	12 to 13
	Tax Assessing	--	--	8 to 12	--	12 to 13
Geodetic Surveys	Basic Control	--	--	9 to 11	10 to 13	12 to 14
	Supplemental Control	--	8 to 9	9 to 12	--	--

TABLE 2.--Continued

Hydraulic and Hydro-logic Development																	
	Canals	--			6 to	9	9 to 10	10 to 12	12 to 14
	Dams	3 to	6	5 to	9	9 to 11	10 to 12	12 to 13
	Drainage Reclamation	--		5 to	9	8 to 11	10 to 12	12 to 13
	Hydroelectric Facilities	2 to	6	5 to	9	8 to 10	10 to 12	12 to 13
	Reservoir	--		5 to	9	8 to 12	10 to 13	12 to 14
	Sewer System	1 to	6	5 to	9	9 to 11	10 to 12	12 to 13
	Water Distribution System	4 to	6	5 to	9	8 to 10	10 to 12	12 to 13
Land Development									
	Agriculture Irrigation	--		5 to	9	8 to 11	10 to 12	12 to 14
	Commercial	3 to	6	5 to	9	8 to 11	10 to 12	12 to 13
	Industrial	1 to	6	5 to	9	8 to 10	10 to 11	12 to 13
	Recreation Parks	--		5 to	9	8 to 11	10 to 12	12 to 13
	Residential	3 to	8	5 to	9	8 to 10	10 to 12	12 to 13
	Wet Lands	--		5 to	9	8 to 12	10 to 13	12 to 15
	Zoning	--		7 to	9	10 to 11	12 to 13	13 to 14
Transportation									
	Aerial Transmission Lines	2 to	5	5 to	9	9 to 11	11 to 12	13 to 14
	Airport Approaches	--		--		--	10 to 12	12 to 14
	Airport Runways	3 to	6	5 to	9	8 to 11	10 to 11	12 to 13
	Airport Terminals	1 to	6	5 to	9	8 to 10	10 to 12	12 to 13
	Highways and Streets	3 to	6	5 to	9	9 to 12	11 to 13	12 to 15
	Mass Transportation	1 to	6	5 to	9	8 to 12	10 to 13	12 to 15
	Railroads	4 to	6	5 to	9	9 to 12	11 to 13	12 to 15
	Secondary, Unimproved, and Temporary Roads	5 to	6	7 to	9	10 to 12	11 to 13	13 to 15
	Tunnels	5 to	6	5 to	9	8 to 12	10 to 13	12 to 15
	Underground Transmission Lines	1 to	6	5 to	9	8 to 11	10 to 12	12 to 13
	Waterfront Facilities[d]	3 to	6	5 to	9	8 to 11	10 to 13	12 to 14

a = Including purchase of rights of way for ownership or easement, whichever is needed.
d = The contours are depth, as well as surface, throughout Waterfront facilities.

TABLE 3-M.

Table 3-M has the descriptive title: Contour Interval in Meters and Scale as a Representative Fraction, by Group, and Land Use. The word 'group' in the title means Map Group Number -- the three words identifying the first column of the table. Table 3-M is the correlator of contour interval and map scale with five principal types of land use. The map group numbers in column one serve as parameters for correlation of contour interval and map scale with the land use types name identified as the separate headings of columns two through six.

Table 3-M was prepared from basic data within the various functional uses of maps, A through E, which are name identified in Column 5 of Table 1. The Map Group Numbers of Table 3-M [also of Table 3-T, hereafter described] enable its user to make map selections which will best fulfill needs by functional use for accomplishing engineering and planning through all of the sequential stages.

TABLE 3-M.   Contour Interval in Meters and Scale as a Representative Fraction, by Group & Land Use.

Map Group Number (1)	Industrial (2)	Urban (3)	Suburban (4)	Farmland (5)	Rangeland (6)
1	0.01 - 0.125 1:25				
2	0.01 - 0.25 1:50 - 1:60	0.01 - 0.4 1:50 - 1:60			
3	0.02 - 0.5 1:100 - 1:200	0.02 - 0.8 1:100 - 1:200			
4	0.04 - 0.75 1:250 - 1:400	0.04 - 1.5 1:250 - 1:400	0.08 - 1.5 1:250 - 1:400		
5	0.06 - 0.8 1:500	0.06 - 1.5 1:500	0.1 - 1.5 1:500	0.1 - 1.5 1:500	
6	0.06 - 1 1:600 - 1:800	0.06 - 2 1:600 - 1:800	0.125 - 2 1:600 - 1:800	0.125 - 2 1:600 - 1:800	0.25 - 4 1:600 - 1:800
7	0.1 - 1.25 1:1,000	0.1 - 2.25 1:1,000	0.15 - 2.25 1:1,000	0.15 - 2.25 1:1,000	0.3 - 4 1:1,000
8	0.1 - 1.5 1:1,250	0.1 - 2.5 1:1,250	0.2 - 2.5 1:1,250	0.2 - 2.5 1:1,250	0.4 - 5 1:1,250
9	0.125 - 2 2,000 - 2,500	0.125 - 4 2,000 - 2,500	0.25 - 4 2,000 - 2,500	0.25 - 4 2,000 - 2,500	0.5 - 8 2,000 - 2,500
10	0.25 - 4 5,000 - 6,000	0.25 - 6 5,000 - 6,000	0.5 - 6 5,000 - 6,000	0.5 - 6 5,000 - 6,000	0.8 - 12.5 5,000 - 6,000
11	0.3 - 5 8,000 - 10,000	0.3 - 10 8,000 - 10,000	0.6 - 10 8,000 - 10,000	0.6 - 10 8,000 - 10,000	1.25 - 20 8,000 - 10,000
12	0.4 - 7.5 12,500 - 20,000	0.4 - 15 12,500 - 20,000	0.8 - 15 12,500 - 20,000	0.8 - 15 12,500 - 20,000	1.5 - 25 12,500 - 20,000
13		0.6 - 20 25,000	1.25 - 20 25,000 - 40,000	1.25 - 20 25,000 - 40,000	2.5 - 40 25,000 - 40,000
14			1.75 - 25 50,000 - 60,000	1.75 - 25 50,000 - 60,000	4 - 50 50,000 - 60,000
15				2.5 - 40 100,000 - 125,000	5 - 80 100,000 - 125,000

EXPLANATION: For each Map Group Number, the range in contour interval in meters is on the first line, and the map scale as a representative fraction is on the second line. For all lines where the [1:] could not be included for lack of space, the numbers are the denominator (N) of the representative fraction expressing map scale. For all tabulated contour intervals and scales, each range indicator is a dash [-] inserted in lieu of the two connective words 'from and to,' so as to keep the tabulation within the width of one page.

TABLE 3-T.

Table 3-T has the descriptive title: Contour Interval in Feet, and Map Scale in Feet per Inch, by Group and Land Use. It is identical to Table 3-M in its structure and intended use. It differs from Table 3-M only in the fact that its contour interval is in feet and its map scale (MS) is in feet per inch. The word 'group' is also a shortening, for lack of space, from the three words--Map Group Number.

TABLE 3-T.   Contour Interval in feet, and Map Scale in Feet per Inch, By Group & Land Use.

Map Group Number	Type of Land Use				
	Industrial	Urban	Suburban	Farmland	Rangeland
(1)	(2)	(3)	(4)	(5)	(6)
1	0.03 - 0.4 2				
2	0.04 - 0.8 4 & 5	0.04 - 1.5 4 & 5			
3	0.08 - 1.5 8 - 16	0.08 - 3 8 - 16			
4	0.1 - 2 20 & 30	0.1 - 4 20 & 30	0.2 - 4 20 & 30		
5	0.2 - 2.5 40	0.2 - 5 40	0.4 - 5 40	0.4 - 5 40	
6	0.2 - 3 50 & 60	0.2 - 6 50 & 60	0.4 - 6 50 & 60	0.4 - 6 50 & 60	0.4 - 10 50 & 60
7	0.3 - 4 80	0.3 - 8 80	0.5 - 8 80	0.5 - 8 80	1 - 12.5 80
8	0.4 - 4 100	0.4 - 8 100	0.6 - 8 100	0.6 - 8 100	1.25 - 15 100
9	0.4 - 6 160 & 200	0.4 - 12.5 160 & 200	0.8 - 12.5 160 & 200	0.8 - 12.5 160 & 200	1.5 - 25 160 & 200
10	0.75 - 10 400	0.75 - 20 400	1.5 - 20 400	1.5 - 20 400	2.5 - 40 400
10	0.8 - 12.5 500	0.8 - 20 500	1.5 - 20 500	1.5 - 20 500	3 - 40 500
11	1 - 12.5 600	1 - 25 600	1.75 - 25 600	1.75 - 25 600	3 - 50 • 600
11	1.25 - 15 800	1.25 - 30 800	2 - 30 800	2 - 30 800	4 - 60 800
12	1.25 -20 1,000 & 1,600	1.25 - 40 1,000 & 1,600	2.5 - 40 1,000 & 1,600	2.5 - 40 1,000 & 1,600	5 - 80 1,000 & 1,600
13		2 - 50 2,000	4 - 50 2,000	4 - 50 2,000	8 - 100 2,000
13			4 - 60 2,500 & 3,000	4 - 60 2,500 & 3,000	8 - 100 2,500 & 3,000
14			5 - 80 4,000	5 - 80 4,000	10 - 150 4,000
14			6 - 80 5,000	6 - 100 5,000	12.5 - 175 5,000
15				8 - 100 8,000	15 - 225 8,000
15				10 125 10,000	20 - 250 10,000

EXPLANATION: For each Map Group Number, the range in contour interval in feet is on the first line, and the map scale in feet per inch is on the second line.

TABLE 4-M.

Table 4-M has the identifying title: Contour Interval Ranges in Meters by Map Use, Group Number, & Representative Fraction. It was developed to correlate the contour interval and map scale for each of five principal types of topography--flat, gently rolling, rolling, hilly, and mountainous--with each of the five identified types of map use letter designated--A, B, C, D, and E. By use of the Map Group Numbers, Table 4-M also serves to correlate the contour interval and map scale ranges with the five types of land use name identified and comprising column headings, numbered (2) through (6), in Table 3-M.

Within Table 4-M, column 1 contains the map use identities A, B, C, D, and E; column 2, the map group numbers from Table 2 in each of their parameter functioning positions; column 3, the map scale number (N) comprising the denominator of the representative fraction expressing map scale, associated with map use and map group number; and column 4 through 18, the minimum, median, and maximum contour interval in meters for the five named types of topography--flat, gently rolling, rolling, hilly, and mountainous.

Each contour interval in meters, as listed in Table 4-M, was computed by set of the first set of fifteen constants identified as 'K' factors in Table 5 in conjunction with the numeral (N) comprising the denominator of the scale as a representative fraction. Each contour interval is rounded to its nearest practicable value for selection and use from the applicable map group number and map scale for fulfilling engineering and/or planning needs according to the type of topography. Such factors can be used to compute the contour interval for any other map scale that might be chosen.

TABLE 4-M.   Contour Interval Ranges in Meters by Map Use, Group Number, & Representative Fraction.

Iden-tity of Map Use (1)	Map Group Num-ber (2)	Scale Numb. (N) 1:(N) (3)	Flat Min. (4)	Flat Med. (5)	Flat Max. (6)	Gently Rolling Min. (7)	Gently Rolling Med. (8)	Gently Rolling Max. (9)	Rolling Min. (10)	Rolling Med. (11)	Rolling Max. (12)	Hilly Min. (13)	Hilly Med. (14)	Hilly Max. (15)	Moun-tainous Min. (16)	Moun-tainous Med. (17)	Moun-tainous Max. (18)
A.	1	25	.01	.02	.04	.02	.04	.06	.04	.06	.125	.06	.125	.25	.125	.25	.5
	2	50	.01	.03	.05	.03	.05	.1	.05	.1	.2	.1	.2	.4	.2	.4	.75
	2	60	.02	.03	.06	.03	.06	.125	.06	.125	.25	.125	.25	.4	.25	.4	.8
	3	100	.02	.04	.08	.04	.08	.15	.08	.15	.3	.15	.3	.6	.3	.6	1
	3	125	.025	.05	.1	.05	.1	.2	.1	.2	.4	.2	.4	.6	.4	.6	1.25
	3	200	.03	.06	.125	.06	.125	.25	.125	.25	.5	.25	.5	.8	.5	.8	1.5
	4	250	.04	.08	.15	.08	.15	.25	.15	.25	.5	.25	.5	1	.5	1	2
	4	400	.05	.1	.2	.1	.2	.4	.2	.4	.75	.4	.75	1.5	.75	1.5	2.5
	5	500	.06	.1	.2	.1	.2	.4	.2	.4	.8	.4	.8	1.5	.8	1.5	3
	6	600	.06	.125	.25	.125	.25	.5	.25	.5	.8	.5	.8	1.75	.8	1.75	4

TABLE 4-M. Continued

B.	5	500	.06	.1	.2	.1	.2	.4	.2	.4	.8	.4	.8	1.5	.8	1.5	3
	6	600	.06	.125	.25	.125	.25	.5	.25	.5	.8	.5	.8	1.75	.8	1.75	4
	6	800	.08	.15	.3	.15	.3	.5	.3	.5	1	.5	1	2	1	2	4
	7	1,000	.1	.15	.3	.15	.3	.6	.3	.6	1.25	.6	1.25	2.25	1.25	2.25	4
	8	1,250	.1	.2	.4	.2	.4	.75	.4	.75	1.5	.75	1.5	2.5	1.5	2.5	5
	9	2,000	.125	.25	.5	.25	.5	1	.5	1	1.75	1	1.75	4	1.75	4	7.5
	9	2,500	.15	.3	.6	.3	.6	1	.6	1	2	1	2	4	2	4	8
C.	8	1,250	.1	.2	.4	.2	.4	.75	.4	.75	1.5	.75	1.5	2.5	1.5	2.5	5
	9	2,000	.125	.25	.5	.25	.5	1	.5	1	1.75	1	1.75	4	1.75	4	7.5
	9	2,500	.15	.3	.6	.3	.6	1	.6	1	2	1	2	4	2	4	8
	10	5,000	.25	.5	.8	.5	.8	1.5	.8	1.5	3	1.5	3	6	3	6	12.5
	10	6,000	.25	.5.	1	.5	1	1.75	1	1.75	4	1.75	4	6	4	6	12.5
	11	8,000	.3	.6	1.25	.6	1.25	2	1.25	2	4	2	4	8	4	8	15
	11	10,000	.4	.6	1.25	.6	1.25	2.5	1.25	2.5	5	2.5	5	10	5	10	20
	12	12,500	.4	.8	1.5	.8	1.5	3	1.5	3	5	3	5	10	5	10	20
D.	10	5,000	.25	.5	.8	.5	.8	1.5	.8	1.5	3	1.5	3	6	3	6	12.5
	10	6,000	.25	.5	1	.5	1	1.75	1	1.75	4	1.75	4	6	4	6	12.5
	11	8,000	.3	.6	1.25	.6	1.25	2	1.25	2	4	2	4	8	4	8	15
	11	10,000	.4	.6	1.25	.6	1.25	2.5	1.25	2.5	5	2.5	5	10	5	10	20
	12	12,500	.4	.8	1.5	.8	1.5	3	1.5	3	5	3	5	10	5	10	20
	12	20,000	.5	1	2	1	2	4	2	4	7.5	4	7.5	15	7.5	15	25
	13	25,000	.6	1.25	2.5	1.25	2.5	4	2.5	4	8	4	8	15	8	15	30
E.	12	12,500	.4	.8	1.5	.8	1.5	3	1.5	3	5	3	5	10	5	10	20
	12	20,000	.5	1	2	1	2	4	2	4	7.5	4	7.5	15	7.5	15	25
	13	25,000	.6	1.25	2.5	1.25	2.5	4	2.5	4	8	4	8	15	8	15	30
	13	30,000	.6	1.25	2.5	1.25	2.5	5	2.5	5	10	5	10	20	10	20	30
	13	40,000	.8	1.5	3	1.5	3	5	3	5	10	5	10	20	10	20	40
	14	50,000	1	1.75	4	1.75	4	6	4	6	12.5	6	12.5	25	12.5	25	50
	14	60,000	1	2	4	2	4	8	4	8	15	8	15	25	15	25	50
	15	100,000	1.5	2.5	5	2.5	5	10	5	10	20	10	20	40	20	40	75
	15	125,000	2	3	6	3	6	12.5	6	12.5	20	12.5	20	40	20	40	80

TABLE 4-T.

Table 4-T has the descriptive title: Contour Interval Ranges in feet by Map Use, Group Number, and Scale in Feet per Inch. It is similar in every aspect to Table 4-M, except the contour interval is in feet, and the map scale (MS) is in feet per inch. The second set of fifteen 'K' factors in Table 5 were used to compute from the map scale (MS) the contour interval in feet for Table 4-T, while effectuating practical rounding. The third set of fifteen 'K' factors in Table 5 in conjunction with the numeral (N), comprising the denominator of map scale as a representative fraction, may be used to compute Table 4-T.

Whatever is subsequently presented by example in use of tables comprising the metric system of measurement in conjunction with Tables 2, 3-M, and 4-M is similarly applicable to the foot system in conjunction with Tables 2, 3-T, and 4-T.

TABLE 4-T.   Contour Interval Ranges in feet by Map Use, Group Number, and Scale in Feet per Inch.

Iden-tity of Map Use	Map Group Num-ber	Map Scale (MS) ft/in.	Flat			Gently Rolling			Rolling			Hilly			Moun-tainous		
			Min.	Med.	Max.	Min.	Med.	Max.	Min.	Med.	Max.	Min.	Med.	Max.	Min.	Med.	Max.
(1)	(2)	(3)	(4)	(5)	(6)	(7)	(8)	(9)	(10)	(11)	(12)	(13)	(14)	(15)	(16)	(17)	(18)
A.	1	2	.03	.06	.1	.06	.1	.2	.1	.2	.4	.2	.4	.8	.4	.8	1.5
	2	4	.04	.08	.15	.08	.15	.3	.15	.3	.6	.3	.6	1.25	.6	1.25	2
	2	5	.06	.1	.2	.1	.2	.4	.2	.4	.8	.4	.8	1.5	.8	1.5	2.5
	3	8	.08	.15	.25	.15	.25	.5	.25	.5	1	.5	1	2	1	2	3
	3	10	.08	.15	.3	.15	.3	.5	.3	.5	1	.5	1	2	1	2	4
	3	16	.1	.2	.4	.2	.4	.8	.4	.8	1.5	.8	1.5	3	1.5	3	5
	4	20	.1	.2	.4	.2	.4	.8	.4	.8	1.5	.8	1.5	3	1.5	3	6
	4	30	.15	.3	.6	.3	.6	1	.6	1	2	1	2	4	2	4	8
	5	40	.2	.4	.6	.4	.6	1.25	.6	1.25	2.5	1.25	2.5	5	2.5	5	10
	6	50	.2	.4	.8	.4	.8	1.5	.8	1.5	3	1.5	3	6	3	6	10
B.	5	40	.2	.4	.6	.4	.6	1.25	.6	1.25	2.5	1.25	2.5	5	2.5	5	10
	6	50	.2	.4	.8	.4	.8	1.5	.8	1.5	3	1.5	3	6	3	6	10
	6	60	.25	.5	.8	.5	.8	1.5	.8	1.5	3	1.5	3	6	3	6	10
	7	80	.3	.5	1	.5	1	2	1	2	4	2	4	8	4	8	12.5
	8	100	.4	.6	1.25	.6	1.25	2.5	1.25	2.5	4	2.5	4	8	4	8	15
	9	160	.4	.8	1.5	.8	1.5	3	1.5	3	6	3	6	10	6	10	20
	9	200	.5	1	1.75	1	1.75	3	1.75	3	6	3	6	12.5	6	12.5	25
C.	8	100	.4	.6	1.25	.6	1.25	2.5	1.25	2.5	4	2.5	4	8	4	8	15
	9	160	.4	.8	1.5	.8	1.5	3	1.5	3	6	3	6	10	6	10	20
	9	200	.5	1	1.75	1	1.75	3	1.75	3	6	3	6	12.5	6	12.5	25
	10	400	.75	1.5	2.5	1.5	2.5	5	2.5	5	10	5	10	20	10	20	40
	10	500	.8	1.5	3	1.5	3	6	3	6	12.5	6	12.5	20	12.5	20	40
	11	600	1	1.75	3	1.75	3	6	3	6	12.5	6	12.5	25	12.5	25	50
	11	800	1.25	2	4	2	4	8	4	8	15	8	15	30	15	30	60
	12	1,000	1.25	2.5	5	2.5	5	10	5	10	20	10	20	30	20	30	60
D.	10	400	.75	1.5	2.5	1.5	2.5	5	2.5	5	10	5	10	20	10	20	40
	10	500	.8	1.5	3	1.5	3	6	3	6	12.5	6	12.5	20	12.5	20	40
	11	600	1	1.75	3	1.75	3	6	3	6	12.5	6	12.5	25	12.5	25	50
	11	800	1.25	2	4	2	4	8	4	8	15	8	15	30	15	30	60
	12	1,000	1.25	2.5	5	2.5	5	10	5	10	20	10	20	30	20	30	60
	12	1,600	1.75	3	6	3	6	12.5	6	12.5	20	12.5	20	40	20	40	80
	13	2,000	2	4	8	4	8	12.5	8	12.5	25	12.5	25	50	25	50	100
E.	12	1,000	1.25	2.5	5	2.5	5	10	5	10	20	10	20	30	20	30	60
	12	1,600	1.75	3	6	3	6	12.5	6	12.5	20	12.5	20	40	20	40	80
	13	2,000	2	4	8	4	8	12.5	8	12.5	25	12.5	25	50	25	50	100
	13	2,500	2.5	4	8	4	8	15	8	15	30	15	30	60	30	60	100
	14	4,000	3	5	10	5	10	20	10	20	40	20	40	80	40	80	150
	14	5,000	4	6	12.5	6	12.5	25	12.5	25	50	25	50	80	50	80	175
	15	8,000	5	8	15	8	15	30	15	30	60	30	60	100	60	100	225
	15	10,000	5	10	20	10	20	40	20	40	60	40	60	125	60	125	250

THE PRACTICABILITY OF EXTREMES IN CONTOUR INTERVAL.

The extremes in contour interval for each map scale are not dominant in the selection and/or design and compilation of topographic maps for engineering and planning purposes. Instead, a contour interval within the vicinity of the median will fulfill most needs associated with the topography, land use, drainage, ground structure, soils, and vegetation within the areas of concern, whether it is small or large.

Nevertheless, with respect to Tables 1, 3-M, 3-T, 4-M, and 4-T, the extremes of small contour interval and large contour interval for each map scale are especially applicable when there are unusual topographic conditions and map use situations. Two examples will be given to illustrate the practicability of the extremes for each map scale within the five tables.

First, it is impracticable for critical design to use a contour interval of 0.06 meter at a map scale (N) of 500, or a contour interval of 0.2 foot at a map scale (MS) of 40 feet per inch, if the topography is mountainous. Such contour intervals, however, would be essential in the use of large map scales if the topography is flat, and the slopes do not exceed 100:1 throughout the area of concern; in such an instance, the closest the contours would be to each other on the map is 12 mm or one-half inch; and for slopes of 200:1 the closest the contours would be to each other on such a map would be 24 mm or one inch. This example could be enlarged upon, but its details portray the effectiveness of the principle.

Second, it would be equally impracticable to use for critical design a contour interval of 3 meters at the map scale (N) of 500, or contour interval of 10 feet at a map scale of 40 feet per inch, if the topography were flat. Such contour intervals, however, would be both practical and effective if the topography were mountainous and the slopes were as steep or steeper than 1:1, in such an instance, the contours would be as close to each other on the topographic map as 6 mm or one-fourth inch for the slopes of 1:1, and closer to each other for all steeper slopes.

Topographic maps have been compiled with a 3 or 4 meter contour interval at the map scale (N) of 500, or with a contour interval of 10 feet at the map scale (MS) of 40 or 50 feet per inch, for the design of bridges, retaining walls, tunnel portals, and so forth within mountainous areas. Also, topographic maps have been compiled with a contour interval of 0.06 or 0.08 meter at the map scale (N) of 500 or 800, or a contour interval of 0.2 or 0.25 foot at a (MS) of 40 or 60 feet per inch for the design of a housing development and its drainage systems on nearly level land within or bordering a city. Otherwise, adequate representation of significant configurations of the surface of the ground would not be attained.

TABLE 5.

Table 5 has the title: K Factors for Computing Contour Interval from Map Scale for Topography. It is the mathematical basis of all presentations within this chapter, and is the foundation for all charts and tables containing contour intervals and map scales. Its usefulness would be limited if it had been prepared without knowledge from experience and the correlation developed from the large number of maps collected, examined, and classified by contour interval, map scale, and map use according to the various types of topography and land use throughout the progressive stages of engineering and planning. This chapter presents a systematic approach for selecting and benefiting from the constants of Table 5 to compute each contour interval from the 0.6 power of the map scale.

Each user of the principles presented herein for Table 5 need not use those constants in the equations. All that needs to be done is to use the charts and other tables. Examples of their use will be presented after introducing effects of the contour-interval-to-map-scale-ratio.

TABLE 5. K Factors for Computing Contour Interval from Map Scale for Topography.

| Map Scale Iden- tity | Type of Contour Interval | Meas- ure- ment Unit | K Factors for each Type of Topography and Map Scale in Feet per Inch and as a Representative Fraction | | | | |
			Flat	Gently Rolling	Rolling	Hilly	Moun- tainous
N	Minimum	Meter	0.0014	0.0027	0.0051	0.0098	0.0187
N	Median	Meter	0.0027	0.0051	0.0098	0.0187	0.0359
N	Maximum	Meter	0.0051	0.0098	0.0187	0.0359	0.0686
MS	Minimum	Foot	0.0204	0.0390	0.0747	0.1428	0.2732
MS	Median	Foot	0.0390	0.0747	0.1428	0.2732	0.5226
MS	Maximum	Foot	0.0747	0.1428	0.2732	0.5226	0.9996
N	Minimum	Foot	0.00459	0.00878	0.01680	0.03213	0.06146
N	Median	Foot	0.00878	0.01680	0.03213	0.06146	0.11758
N	Maximum	Foot	0.01680	0.03213	0.06146	0.11758	0.22492

EXPLANATION: N [which may be called the scale number] is the numeral comprising the denominator of the representative fraction expressing map scale, and MS is the map scale in feet per inch. Any K Factor, except the first, within each of the three different Map Scale Identity sets is progressively 1.91294 times larger than its preceding value within that set. For computing each contour interval, CI, in meters or in feet, select the appropriate K factor from Table 5 for multiplication by the 0.6 power of N or MS, respectively, in one of these equations:

$$CI = K \, (N)^{0.6} = \text{Meters or Feet}. \qquad CI = K \, (MS)^{0.6} = \text{Feet}.$$

CHART 3.

Like Charts 1 and 2, Chart 3 is also logarithmic. Its abscissa is the contour interval in meters and its ordinate is the denominator (N) of the representative fraction expressing map scale. In addition to three sloping lines identifying the minimum, median, and maximum contour interval, there are sixteen other sloping lines on Chart 3. Each line is composed of a continuous series of points. Each point identifies the ratio resulting from the contour interval it represents in meters being divided by denominator (N) of the scale of the map on which it is or will be delineated. The ratio-of-contour-interval-to-map-scale chart thereby portrays numerous ratios for contour interval selection by map scale to fulfill requirements on a specific engineering project. The contour-interval-to-map-scale ratio is written 1/100 for the largest ratio and 1/100,000 for the smallest ratio on the chart, with other ratios lying between those extremes. A line could not be placed on the chart for every contour-interval-to-map-scale (N) ratio that will occur in practice. [Such ratios should not be confused with map scale expressed as a representative fraction in this manner 1:1,000.]

The contour-interval-to-map-scale ratios depicted on Chart 3 represent numerical comparisons when the contour interval is in meter(s) and the map scale is expressed as a representative fraction. Each of such ratios permits identifying whether or not a topographic map can be compiled solely by the most commonly used current photogrammetric methods, regardless of the photographic visibility of the ground from the aerial camera position in its transporting aircraft.

Each user of Chart 3 can visually interpolate lines applicable to a particular map from the ratios for which lines have been drawn on the chart. For example, if the contour interval is 5 meters and the denominator (N) of the map scale is 10,000, the ratio of contour interval to that scale is 1/2,000; and if the contour interval is 0.1 meter and the denominator (N) of the map scale is 200, the ratio of contour interval to that map scale is also 1/2,000.

If the required topographic map, in which the contour interval is in meters and the map scale is expressed as a representative fraction, has a ratio of contour interval to denominator (N) of map scale which is smaller than 1/1,600 or 1/2,000, and only the most commonly used photogrammetric methods are available, the map would have to be compiled from surveys on the ground. If that ratio is equal to 1/1,600 or 1/2,000 or larger, and if the ground is adequately visible from the aerial camera position in its transporting aircraft, the topographic map may be compiled photogrammetrically and be expected to comply readily with specified accuracies.

For all cross sections and/or spot elevations measured by similar photogrammetric methods for use in conjunction with maps [as if the maps were to have contours, although they do not] the analogous ratio in vertical accuracy in meters to the number (N) comprising the denominator of the map scale required for the cross sections and spot elevations should have a ratio in meters to the denominator (N) of map scale which is no smaller than 1/8,000 or 1/10,000, if it is expected that each

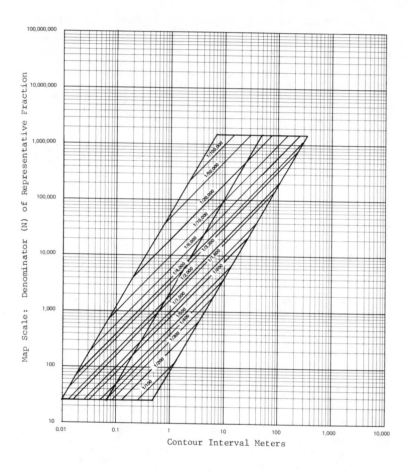

Chart 3.   Ratio of Contour Interval to Map Scale

(The Labeled lines of 1/100 to 1/100,000 are meters to denominators of
representative fractions.)

measurement will comply with specification accuracies. In a numerical sense, for example, this means that if the vertical accuracy desired in the measured cross sections and/or spot elevations is to be 0.2 meter, the scale at which the map is being compiled while the cross sections and/or spot elevations are being measured on the stereoscopic model in the photogrammetric instrument divided by that desired accuracy should result in a ratio no smaller than 1/8,000 or 1/10,000. Consequently, the scale of the map for which such vertical measurements are made would be 1:1,600 or 1:2,000. If contours were measured and delineated photogrammetrically from the same stereoscopic model, however, their interval could be one meter, their accuracy should comply with the specifications for contours, and the contour-interval-to-map-scale ratio would be no smaller than 1/1,600 or 1/2,000, because the map scale would be 1:1,600 or 1:2,000, respectively. For all other contour intervals, as well as for accuracy of measurement for cross sections and/or spot elevations in meters, such ratios are equally applicable.

Additional information will be given regarding the contour-interval-to map-scale ratios usable for ascertaining the feasibility of using photogrammetric methods of measuring and mapping after presenting Chart 4, because the effects of such ratios are not 'fine-line' restricting. They become variable, although they do have a practical limit, depending upon the status of technology and practice, and the capability of the photogrammetric instruments and/or system and how well they are employed.

CHART 4

Like Chart 3, Chart 4 is logarithmic, but its abscissa is the contour interval in feet, and its ordinate is the map scale (MS) in feet per inch. Also, in addition to the three sloping lines identifying the minimum, median, and maximum contour interval for each map scale, there are other sloping lines which portray the ratio of contour interval in feet to the map scale (MS) in feet per inch.

The contour-interval-to-map-scale ratio is written 1/2 for the largest and 1/12,500 for the smallest ratio on the chart, with other ratios lying between those extremes. The contour-interval-to-map-scale ratios within Chart 4 comprise numerical comparisons only for the foot system of measurements.

If a topographic map has a contour-interval-to-map-scale ratio at the compilation scale smaller that 1/40 or 1/50, the map should be compiled from surveys on the ground. Likewise, any cross sections and spot elevations to be used in conjunction with maps [as if the maps were to have contours, although they do not] the analogous ratio in the vertical accuracy in feet to the map scale in feet per inch required for the cross sections and/or spot elevations measured by the photogrammetric methods identified in the previous paragraph must be not smaller than 1/200 or 1/250, or the measurements will usually not comply with specified accuracies. For example, in a numerical sense, this means that if the vertical accuracy desired in the measured cross sections and/or spot elevations is to be 0.2 foot, in lieu of compiling a topographic map with a 1-foot contour interval, the compilation scale of the map should be not smaller than 40 or 50 feet per inch.

Accordingly, for 0.2 foot vertical accuracy in each measured cross section and/or spot elevation, the photogrammetric measurement scale cannot be 80, 100, 200, or smaller scale in feet per inch; but, it must be 40 or 50 feet per inch. To be more certain of achieving such accuracy in this instance, the scale for both photogrammetric compilation of maps and measurements of cross sections and/or spot elevations in conjunction therewith might be 30, 25, or 20 feet per inch.

Another example is cited. The contour interval is 5 feet, the map scale is 200 feet per inch; or the required vertical accuracy is one foot for the photogrammetrically measured cross sections and/or spot elevations in lieu of contours. Thus, the ratio of contour-interval-to-map-scale is 1/40, and the ratio of vertical accuracy in feet to the map scale in feet per inch is 1/200 for the cross sections and/or spot elevations.

Such relationships mean, if the ground cover is not too dense and tall, that the topographic map with its 5-foot contour interval can be compiled photogrammetrically at the scale of 200 feet per inch, or that other maps of similar scale can be compiled while the cross sections and/or spot elevations are also measured photogrammetrically, with likelihood of their complying with a 1-foot accuracy specification.

Thus, if the contour-interval-to-map-scale ratio at the compilation scale is smaller than 1/40 or 1/50, such as 1/80, 1/100, and so on, maps compiled photogrammetrically having contour intervals and scales which result in similar or smaller ratios cannnot be expected to comply with accuracy specifications for the contours. Likewise, if such ratios are smaller that 1/200 or 1/250 for the intended accuracy of cross sections and/or spot elevations measured photogrammetrically at the map compilation scale, such measurements cannot be expected to fulfill their vertical accuracy specifications.

The reason for such limiting ratios is the fact that contours delineated or the cross sections and spot elevations measured by the most commonly used photogrammetric methods will not usually comply with accuracies required for engineering purposes unless the ratio of contour-interval-to-flight-height is not smaller than 1/1,200 for contours and not smaller than 1/6,000 in ratio of accuracy to flight-height for cross sections and spot elevations, respectively. This is so regardless of the unit of linear measurement--a meter or a foot.

Use of the contour-interval-to-map-scale ratio comprises an indirect way of considering the effect of the contour-interval-to-flight-height ratio (usually termed the "C"-factor) without having to compute that ratio for the compilation scale of every topographic map or other maps for which cross sections and/or spot elevations are to be measured before deciding whether or not photogrammetric methods will suffice.

In topographic map compilation cases, whenever the contour interval is in meters and the map scale is expressed as a representative fraction, use of Chart 3 will obviate need for ascertaining the contour-interval-to-flight-height ratio. The reason for this is that each ratio of contour interval to the denominator (N) of map scale presented on Chart 3 serves in an indirect way for obviating need to ascertain each ratio of contour interval in meters to flight-height in meters. Likewise, when the contour interval is in feet and the map scale is in feet per inch, use of Chart 4 will obviate need for ascertaining the contour-interval-to-flight-height ratio.

Thereby, there is no need to determine the contour-interval-to-flight-height ratio for each topographic map to ascertain if it can be compiled photogrammetrically, or to determine whether or not cross sections and/or spot elevations may also be measured photogrammetrically for use with another type of map. Merely use the appropriate chart: Chart 3 for the metric system and Chart 4 for the foot system of measurement.

It is recognized that measurements can be made more accurately by photogrammetric methods than are illustrated within the preceding paragraphs. This is especially so if repetitive measurements are made and their average is recorded. In actual practice, however, the measuring mark is in continuous motion when contours are being delineated. In that mode, the mark may be in any one of three positional conditions: (1) exactly on the ground, (2) slightly floating, or (3) somewhat digging with respect to the surface of the steroscopic

45

model being viewed. Hence, the reason for making guidance presentations herein which have greater assurance of attaining the desired operational results than are likely if they are ignored.

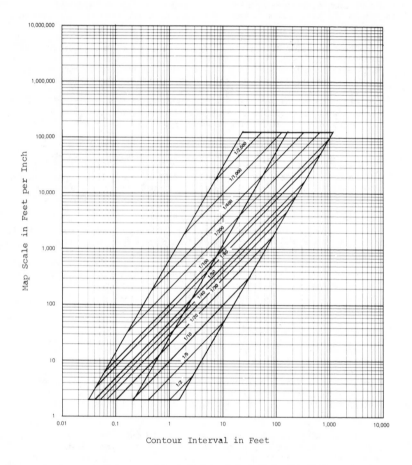

Chart 4.   Ratio of Contour Interval to Map Scale

(The labeled lines of 1/12 to 1/12,500 are Feet to Feet per Inch.)

EFFECTS OF IMPROVEMENT IN PHOTOGRAMMETRIC TECHNOLOGY AND INSTRUMENTATION

An important aspect of the Charts 3 and 4 is that technology and instrumentation for making measurements and compiling maps photogrammetrically are not static. Progress is being made continually in improving the accuracy and reliability capabilities of existing instruments and/or systems. In addition, new techniques and instrumental approaches are coming into being on a sustained basis which offer improvements in the accuracy and reliability of measurements and map compilations which can be done photogrammetrically.

The functional examples herein apply to utilization of a specific generation of photogrammetric techniques and instrumentation. What may currently be the most generally used instruments may be given the rating of one. Another generation of instruments may have the rating of two, and a still later generation the rating of three and so forth. Such ratings have no distinction except to serve as a means of identifying progressively, by number: (1) The functioning capabilities of photogrammetric techniques, procedures, and instruments now in somewhat general use, (2) those capabilities that are based upon improvements and/or new developments that recently became available and are in selective use, and (3) instrumentation techniques and systems that will be developed and become economically available and usable in the future.

When and where improvements occur in photogrammetric technology and instrumentation, and any one or more of them are employed, their advantage is in extending the functional effect of the ratio of contour interval in meters to the denominator (N) of the map scale as a representative fraction from 1/1,600 or 1/2,000 to 1/2,400, 1/3,200, or 1/4,000, and so on, as feasible. The same is true for the foot system of measurement, wherein the ratio of contour interval in feet to the map scale (MS) in feet per inch is extended from the ratio of 1/40 or 1/50 to 1/60, 1/80, 1/100, and so on.

Selection and factual use of any one of such smaller ratios is governed by the quality and precision in performance of the photogrammetric instruments or system, and in reliability of the accuracy of measurements made and maps compiled thereby.

Such extensions are equally applicable to change in the ratio of the vertical accuracy in cross section and spot elevation measurements to map scale as they are for the change in the ratio of contour-interval-to-map-scale, as previously illustrated. Therefore, it is possible for the ratio of vertical accuracy in cross section and spot elevation measurements to map scale to change from the ratios of 1/8,000 or 1/10,000 to 1/12,000, 1/16,000, or 1/20,000, and so on, for the metric system; and from ratios of 1/200 or 1/250 to 1/300, 1/400, 1/500, and so on, for the foot system; the possible change depends on the operational reliability and accuracy of the photogrammetric methods available, and how well they are used.

47

Qualified professional engineers with functioning knowledge and experience in photogrammetric technology and instrumentation or certified photogrammetrists should be consulted in order to ascertain knowledgeably the specific functional effects of photogrammetric technology, instruments, and/or systems. Thereby, there is more assurance of attaining the most effective, reliable, and economical results from photogrammetric methods of measurement and map compilation, while making realistically feasible use of the ratios of contour-interval-to-map-scale, and the ratios of vertical accuracy in cross sections and spot elevations to map scale.

EXAMPLES.

The subsequent examples are selective. One example cannot be representative of all possible uses of topographic and other maps. The examples are as indicative as possible of an effective way to select maps by contour interval and/or scale to fulfill engineering and planning requirements and serve well, according to the stage of use of each map, for the project of concern within every one of the major fields of activity.

Every example is introduced by five criteria. Each criterion has its influence in determining the contour interval of each topographic map, the scale of all maps, and the accuracy to be expected in measurement of cross sections and spot elevations.

The meaning and significance of each of the five criteria utilized are as follows:

1.   PROJECT: The specific construction or other type of activity for which the engineering and associated work is to be undertaken. In most instances the project identity may be similar to or exactly the same by name as one of the listings within the first two columns of Table 2 under the heading "Fields of Activity."

2.   IDENTITY OF LAND USE: Land use within and/or throughout the area for which maps are required and surveys will have to be made for accomplishing all subsequent engineering and/or associated work. A typical word description is given of five types of significant land use, comprising the separate headings of columns 2 through 6 of Tables 3-M and 3-T.

3.   PRINCIPAL TOPOGRAPHY: This criterion pertains to defining the type of topography by use of one or more of the descriptive words identifying the five headings within Tables 4-M and 4-T--columns 4 to 6, 7 to 9, 10 to 12, 13 to 15, and 16 to 18, respectively, whichever represent(s) the principal type(s) of topography throughout the project area.

4.   CHARACTERIZATION: Each project will have characterization, which, in essence, governs the intricacy required in achieving adequate representation of all topographic, land use, and other principal details required on the maps. Characterization may be

48

represented by any of three levels of detail, as one, two, or three.

As an example, if the topography is gently rolling to rolling, the contour interval for one project within such a topographic area might need to be one-half the size of the contour interval that would be adequate for another type of project. Within an area comprising such topography, general design of an industrial complex may need a contour interval as small as 0.25 meter; whereas, general design of a highway could be accomplished with a contour interval of 0.5 meter. Accordingly, in this example, the level of topographic configuration by contours for the industrial complex would be one and for the highway it would be two.

To further review the application of this principle by specific example, refer to Table 4-M. Opposite the Identity of Map Use B, Column 1; the first two Map Group Numbers 6, Column 2; and the Scale Number 600, Column 3; compare the Contour Interval in Columns 8 and 10 with the contour interval in Columns 9 and 11 for Gently Rolling and Rolling Topography.

It is evident these intervals pertain to the median and minimum and to maximum and median, respectively, for the two different types of topography. Obviously, the 0.25-meter contour interval will provide more topographic representations than it is possible to attain with the 0.5 meter contour interval. This is so although all contours of either interval may have equality with respect to delineation of topographic configurations and elevation accuracy throughout the mapped area.

Therefore, as previously indicated for these two specific examples, the characterization level is one for the industrial complex and it is two for the highway.

5. MAP-COMPILATION FACTOR: Each map-compilation factor is merely the numeral one, two, or three. It is a distinguishing representation of the recognized level of accuracy and reliability of the technology and instrumentation which it is expected will be employed [and will be accepted] for doing the measuring and mapping by photogrammetric methods. Accordingly, each factor is not static--the numeral one representing the least accurate, two the intermediate, and three the most accurate photogrammetric instrumentation system.

Each map-compilation factor changes in its functional application as improvements become economically feasible and acceptable, and are used. Thus, the consultation of qualified professional engineers knowledgeable and experienced in photogrammetry or certified photogrammetrists is advisable before undertaking any measuring or map compilation by photogrammetric methods.

Example 1.

Criteria: 1. Project--Improving an existing subdivision.

2. Identity of land use--An abandoned subdivision.

3. Principal Topography--Gently rolling to rolling.

4. Characterization--Level one for general configurations of the topography.

5. Map-Compilation Factor--One, two, or three, whichever is photogrammetrically feasible.

Such criteria establish the project as Residential within Land Development and the use classification is Suburban. Topographic map selection begins with the basic Table 1. Employing the meter instead of the foot as the distance measuring unit, the principles and procedures of contour interval and map scale selection are applied in these progressive steps for this project:

1. Refer to Table 1. Keep in mind that the site is already known. Thus, Survey of Area and Survey of Alternatives are not required. Any planning, other than Micro (map use C) is not necessary. General Design (functional classification B for the maps) would be essential and adequate for preparation of detailed plans to accomplish the improvement.

2. Use Table 2. It contains the Map Group Numbers correlating Contour Interval and Map Scale for the various types of engineering projects. Examine this table as the first step in selecting the topographic map which will best serve under the governing criteria for each of the two different functional uses of maps for this project. The criteria of Example 1 make the project classification land development in a residential area. Thus, in Table 2, the map group numbers pertaining to functional use of maps for Micro Planning, C, of Table 1, are 8 to 10; and for General Design, B, they are 5 to 9.

3. Refer to Table 3-M. Use column 4 for Suburban land use. Inasmuch as this is not a complicated or difficult planning task, use Map Group Number 10 in Column 1. It is the largest map group number selected for C, Micro Planning, from opposite Residential within the classification of Land Development in Table 2. Again refer to Column 4, opposite Map Group Number 10 in Table 3-M. It will be seen that the contour interval range is 0.5 to 6 meters and the map scale Numbers are 5,000 and 6,000.

4. Refer to Table 4-M. For Map Use C in column 1 and Map Group Number 10 of Column 2, select the scale Number 5,000, Column 3, the larger of the two map scales, and the contour interval of 1.5 meters, Columns 9 and 11, the maximum of Gently Rolling and the median for Rolling topography.

This selection is made because the characterization level is one. If it had been two, an 0.8-meter contour interval would have been selected, which is the median for gently rolling and minimum for rolling topography, respectively, Columns 8 and 10.

[Explanation: Although the 1.75-meter contour interval and the map scale Number of 6,000 would be adequate for accomplishing micro planning for this residential project having the characterization level of one, the 1.5-meter contour interval is selected because it is more commonly used than the 1.75 meter. Thus, selection of the 1.5-meter contour interval within Map Group 10 results in selection of the map scale Number 5,000 instead of 6,000. Personal preference, however, may cause use of hard conversion, which would result in use of a larger contour interval, such as 2 meters, in conjunction with the scale Number 6,000, and in such a case the characterization level would be reduced to less than one.]

 5. Refer to Table 1. The design, as determined in step 1, would not be other than general in character. Thus, the functional map use identity would be B from Table 1.

Use Table 2. Refer to Map Use B opposite the classification land development and the project in a residential area. Select Map Group Number 7, which is a median within the range of Map Group Numbers 5 to 9.

 6. Use Table 3-M. For suburban land use, Column 4, opposite Map Group Number 7 in Column 1, select the contour interval range of 0.15 to 2.25 meters, and the map scale of 1:1,000.

 7. Use Table 4-M. For map use B, Column 1, and Map Group Number 7, Column 2, select the contour interval 0.6 meter from Columns 9 and 11, which is the maximum for Gently Rolling and median for Rolling topography with the map scale Number 1,000, Column 3. As for the Micro Planning, the characterization level is also one for General Design. Accordingly, the contour interval of 0.3 meter, Columns 8 and 10, was not selected, but it would have been if the characterization level had been two for this project.

[Second Explanation: There may be preference for hard conversion, and if this is so, each user of the charts and tables functionally presented herein for use of the metric system of measurement may employ such conversion. The effects, however, of upward or downward application of hard conversion, whichever is used, must be acknowledged, not ignored. To change the contour interval from 0.6 meter to 1 meter on the maps which will be used for design of this project would have the effect of making its characterization level much less than one. In contrast, a change from 0.6 meter to 0.5 meter would be the same as changing its characterization level toward two. Moreover the significance of such conversions should be recognized and accounted for when and wherever photogrammetric methods will be employed for compiling the required topographic maps.]

Such procedure has made full application of the four tables

designed and presented for sequential use. All aspects of the engineering and planning work to be done have been considered. The contour interval and scale for the topographic maps will be adequate for fulfilling requirements in each functionally successive stage of Micro planning and Design for improving the existing subdivision.

The 1.5-meter contour interval and the map scale of 1:5,000 selected for the Micro Planning presents a contour-interval-to-map-scale ratio smaller than 1/3,200 (Chart 3). Accordingly, if the map-compilation factor is limited to one, this map should not be compiled by photogrammetric methods, even though the vegetative ground cover may not be dense and tall.

Under similar vegetative ground cover conditions, if the map-compilation factor of two can be employed, the contour-interval-to-map-scale ratio of 1/3,200 will admit the use of photogrammetric methods with an expectation that the planimetric details and contours will comply adequately with accuracy specifications.

The 0.6-meter contour interval and the map scale of 1:1,000 selected for the General Design presents a truncated contour-interval-to-map-scale ratio of 1/1,600 (Chart 3). Thus, with a map-compilation factor of one, these maps may be compiled easily by photogrammetric methods, provided the vegetative ground cover of the area of survey is not too dense and tall.

Example 2.

Criteria: 1. Project—Extending the existing runway of an airport.

2. Present land use—High class subdivision surrounding the airport.

3. Topography—Flat.

4. Characterization—Level one for specific configurations and two for general configurations of the topography.

5. Factor of Compilation—One, two, or three, whichever is photogrammetrically feasible.

Such criteria classify the area Suburban, and the project Airport Runways under the classification of Transportation in Table 2. Using the foot instead of the meter as the distance measurement unit, the steps in selection of topographic maps for use in the two engineering stages of work for this airport runway project are:

1. Refer to Table 1. The site is known and all stages of map use preceding site selection are unnecessary. It is evident the two functional map uses for this project are C, Micro Planning; and A, Critical Design.

2-C. Use Table 2. To accomplish the Micro Planning, C, for this project, select opposite Airport Runways the Map Group Number 9, which can be considered the smaller median within the range of 8 to 11.

2-A. Use Table 2. For Critical Design, A, select opposite Airport Runways Map Group Number 5, which may be thought of as the larger median within the range of 3 to 6 for this project.

3-C. Use Table 3-T. Opposite Map Group Number 9, Column 1, and in Column 4 for suburban land use, note the contour interval range of 0.8 foot to 12.5 feet, and the map scales of 160 and 200 feet per inch.

3-A. Use Table 3-T. Opposite Map Group Number 5, Column 1, choose from Column 4 the contour interval range of 0.4 foot to 5 feet and the map scale of 40 feet per inch.

4-C. Use Table 4-T. For map use C and opposite Map Group Number 9, Column 2, select the map scale of 200 feet per inch, Column 3, and the contour interval of 1 foot, Column 5, which is the median for flat topography. This contour interval selection is made because the Characterization level is two for general configurations of the topography.

53

4-A. Use Table 4-T. For map use A, Column 1, opposite map group 5, Column 2, and the map scale of 40 feet per inch, Column 3, select the 0.2- foot contour interval, Column 4, which is the minimum for flat topography. Do this because the Characterization level is one to get specific configurations of the topography.

[Explanation: The contour interval selected for both functional map uses, C, and A, fulfills the project Characterization levels of one and two, respectively, in their reverse order of use for this airport runway project. Had the project Characterization level been limited to one, the contour interval selection would have been 0.5 foot, Column 4, for the functional map use, C, instead of 1 foot, Column 5.]

The resultant ratio of contour-interval-to-map-scale for the two different topographic maps [1 foot at 200 feet per inch, and 0.2 foot at 40 feet per inch] would be 1/200 for map use C, and 1/200 for map use A, respectively. Both are smaller than the limit of 1/40 or 1/50, which are consonant with the map-compilation factor of one. Therefore, all topographic maps for each of the two engineering stages for this airport runway extension project would have to be compiled from surveys made on the ground. Otherwise, if photogrammetric methods are to be employed, such methods would have to admit use of a map-compilation factor which is three in compiling maps for map uses C, and B, if the maps are to be expected to comply with accuracy specifications.

Example 3.

Criteria: 1. Project—Property survey for land acquisition to construct an over-the-horizon communication system.

2. Present land use—Farmland.

3. Topography—Hilly.

4. Characterization—Level one for General configurations in the planimetry.

5. Map-Compilation Factor—One.

Using the metric instead of the foot system of measurement, steps for selecting the topographic map needed from this one stage land acquisition surveying project are:

1. Refer to Table 1. The criteria for and the critical alinement and location restrictions associated with this project, and the fact that its site is already known, limit map needs to one engineering stage, which is B in the General Design classification.

2. Use Table 2. Opposite Land Acquisition within the classification Cadastral Operations for map use B, select Map Group Number 6, which is one number less than a median within the range of 5 to 9.

54

3. Use Table 3-M. For Map Group Number 6, Column 1, select from Column 5 the largest scale 1:600, instead of 1:800, for Farmland.

4. For land acquisition, contours are not usually essential. Thus, Table 4-M need not be used for map use B, Column 1. But, if contours are desired, choose from opposite Map Group Number 6, Column 2, a contour interval of 0.8 or 1 meter, according to preference, as these intervals are the median, Column 14, for hilly topography, and the map scale Numbers 600 and 800, respectively, Column 3. Inasmuch as the Characterization level is one, however, the 1-meter contour interval should not be too large at the map scale of 1:600; and it could be selected in preference to the 0.8-meter interval in order to reduce the costs of map compilation.

If contours are desired on each land-acquisition map for this project, the ratio of contour-interval-to-map-scale of 1:600 would be 1/750 for the 0.8 meter interval and 1/600 for the 1-meter interval. Thus, with a map-compilation factor of one, either map could be compiled photogrammetrically with expectations of easily fulfilling accuracy specifications. If contours are not included, that ratio is of no consequence for this particular project in determining whether or not photogrammetric methods of compilation can be used effectively. The limitations caused by vegetative ground cover, and accuracy required in horizontal positioning on the maps of essential topographic, land use, and property boundary details would govern.

Example 4.

Criteria: 1. Project—Determination of feasible route alternatives, selection of a route, survey of selected route, for design of the highway, preparation of detailed construction plans, and location survey and construction surveys for the highway.

2. Land use—From suburban through farm land and range land, and back to suburban.

3. Topography—Rolling to hilly.

4. Characterization—Level one for General configurations of the topography.

5. Map-Compilation Factor—one, two, or three, whichever is photogrammetrically feasible.

From the criteria it is evident that this is a Highway engineering project under the classification of Transportation. Topographic maps will be used throughout all essential stages from survey of area to design and subsequent location survey and construction surveys.

Using the foot instead of the metric system of measurement, steps in selection of topographic maps for acquisition from available sources, if available, and for design and compilation of maps, as necessary, for use in the successive engineering stages of this project are:

1.   Refer to Table 1. The first engineering stage is to determine feasible route alternatives. Consequently, a Survey of Area must be undertaken and completed, which is sometimes thought of as regional planning, and the functional use of maps has the E classification.

2.   Use Table 2. Select from opposite Highways and Streets, within the Transportation classification, from Column E the Map Group Number 13, which is almost a median between the numbers 12 and 15.

3.   Use Table 3-T. Opposite Map Group Number 13, Column 1, select the contour interval range of 4 feet, minimum for Suburban land use, Column 4, to 100 feet, maximum for Rangeland, Column 6, and the map scale range of from 2,000 to 3,000 feet per inch, Columns 4 to 6.

4.   Use Table 4-T. For map use E, Column 1, select opposite Map Group Number 13, Column 2, the map scale of 2,000 feet per inch, Column 3, and the contour interval of 25 feet, Columns 12 and 14, which is the maximum for rolling and the median for hilly topography, because the Characterization level is one.

Topographic maps having the required 25-foot contour interval at the scale of 2,000 feet per inch may usually be obtained from available sources. If such is the case and if the maps were compiled by photogrammetric methods, their initial compilation scale ought to have been as large or larger than 1,250 feet per inch before they were reduced to the scale of 2,000 feet per inch if the instrumentation system were limited to the 1/50 contour-interval-to-map-scale ratio, meaning a map-compilation factor of one.

Also, if the topographic maps are not obtainable from an available source, and if photogrammetric methods or compilation are to be used, their contour-interval-to-map-scale ratio of 1/80 will require use of a map-compilation factor of two, otherwise, if a map-compilation factor of one has to be used to maintain a 1/40 contour-interval-to-map-scale ratio, compilation would be accomplished at the scale of 1,000 feet per inch, and thereafter each map would be reduced photographically to the 2,000-feet-per-inch scale.

5.   Refer to Table 1. For the Survey of Route Alternatives, select the functional map use classification D, identified as local planning. Do this because the project Characterization level is one.

6.   Refer to Table 2. Select from opposite Highways and Streets, within the classification of Transportation, Map Group Number 12, the middle number between numbers 11 to 13, Column D.

7.   Use Table 3-T.   Opposite Map Group Number 12, Column 1,
select the contour interval range of 2.5 feet, minimum for suburban
land use, to 80 feet, maximum for rangeland, and the map scale
range of 1,000 to 1,600 feet per inch, Columns 4 to 6,
respectively.

8.   Use Table 4-T.   For map use D, Column 1, select opposite Map
Group Number 12, Column 2, the map scale of 1,000 feet per inch,
Column 3, and the contour interval 10 feet, Columns 11 and 13,
which is the median for rolling and the minimum for hilly
topography.

The contour interval of 10 feet and the map scale of 1,000 feet per
inch for survey of the route alternatives in this stage comprises a
contour-interval-to-map-scale ratio of 1/100.   Accordingly, if
photogrammetric methods of compilation are to be employed, the
photogrammetric system must fulfill the requirements of a map-
compilation factor of two, or preferably three, before the maps will
comply with accuracy specifications.

To attain a contour-interval-to-map-scale ratio of 1/40 or 1/50 by
use of a map-compilation factor of one, topographic map compilation at
the scale of 400 or 500 feet per inch and then photographic reduction to
the scale of 1,000 feet per inch would be necessary to achieve
compliance with accuracy specifications.

Moreover, if these required maps can be obtained from an available
source, their initial compilation scale ought to have been as large or
larger than 400 to 500 feet per inch if a map-compilation factor of one
applied to the photogrammetric instrumentation system that had been
used.   Otherwise, they may not conform to the accuracy specifications.

9.   Refer to Table 1.   A Survey of Design will have to be made of
the selected route to compile the topographic maps required for
designing the highway location on that route and for preparation of
detailed construction plans.   First, select functional map use B
for General Design.

10.   Use Table 2.   Opposite Highways and Streets within the
classification of Transportation, select for Functional Map Use B
the Map Group Number 8, which is next to the largest of the 5 to 9
Map Group Numbers.   Do this because the land use and topography are
not difficult or too restrictive.

11.   Use Table 3-T.   Opposite Map Group Number 8, Column 1, select
from Column 4 for Suburban land use the minimum 0.6 foot contour
interval, and from Column 6 for Rangeland the maximum 15 foot
contour interval.   The applicable map scale is 100 feet per inch
for each of the three types of land use existing between the
terminal points.

12.   Use Table 4-T.   Within the functional map use classification
B, Column 1, opposite Map Group Number 8, Column 2, and the scale
of 100 feet per inch, Column 3, select the contour interval of 2.5

feet, Columns 11 and 13. It is the median for rolling and minimum for hilly topography.

Topographic maps having the 2.5-foot contour interval and the 100-feet-per-inch scale will be adequate for accomplishing the design, for preparing the detailed construction plans, for making the location survey, and for doing all essential construction surveys.

The contour-interval-to-map-scale ratio of 1/40 on all topographic maps with the 2.5-foot contour interval at the scale of 100 feet per inch which will be used for General Design on this project will readily admit use of a map-compilation factor of one in utilization of photogrammetric methods. Thus, only tall and dense vegetation would preclude using a photogrammetric system of mapping to compile maps for use in accomplishing the General Design for this project.

Topographic maps having a smaller contour interval at a larger scale would be required at sites along the selected route for interchanges; bridges over large streams and rivers, and under or over other highways and railroads; and at sites where other intricate structures are required, and for which Critical Design is essential.

13.   Refer to Table 1. Choose Critical Design, A, for survey to compile maps of specific sites wherever the topography and design are critical within the selected route.

14.   Use Table 2. Select from opposite Highways and Streets under the classification Transportation for functional map use A the Map Group Number 4 from within the Map Group Numbers 3 to 6.

15.   Use Table 3-T. Opposite Map Group Number 4, Column 1, choose from Column 4 the contour interval range of 0.2 foot to 4 feet and the map scales of 20 and 30 feet per inch.

16.   Use Table 4-T. For functional map use A, Column 1, and Map Group Number 4, Column 2, select the map scale of 20 feet per inch, Column 3, and choose a contour interval of 1 foot, which is 0.2 foot larger than the 0.8 of a foot interval, columns 11 and 13, comprising the median for Rolling and the minimum Hilly topography.

If the designer has preference for the scale of 25 or 30 feet per inch, either scale can be used in conjunction with the 1-foot contour interval, in lieu of the scale of 20 feet per inch, for all critical design sites along the selected route of this highway engineering project.

Also, if the designer has preference for the 0.8-foot contour interval, instead of the 1-foot interval, there would be no practical reason for not so doing. The 1-foot contour interval is adequate for the Characterization level one for this project. Compilation costs would be increased, of course, by requiring the 0.8-foot contour interval instead of the 1-foot.

For such topographic maps of sites for which critical design is required, the contour-interval-to-map-scale ratio is 1/20. Accordingly, if aerial photographs can be taken at large enough scale, a map-compilation factor of one can be employed. But, if the flight height which has to be used limits the scale of the aerial photography, which would usually be of the vertical type, to a scale smaller than the photogrammetric instrumentation system would accept for map compilation at the scale of 20 feet per inch, then another, more accurate, system would have to be available and employed to which the map-compilation factor of two or of three is applicable. Otherwise, map compilation by photogrammetric methods would have to be made at a smaller scale and the maps subsequently enlarged photographically to the scale of 20 feet per inch.

If the enlargement technique is employed, the user of the photogrammetric system, as well as the user of the resultant topographic maps, should be sure the maps fulfill accuracy requirements. In such cases, the consultation procedure should be employed before proceeding with use of a photogrammetric system of topographic map compilation. Otherwise, the required topographic maps should be compiled from surveys made on the ground for all sites where critical design is required. In either case, surveys should be made to ascertain the horizontal position and elevational accuracy of all details on the maps before they are used. (See Chapter 4).

Example 5.

Criteria: 1. Project—Determination of feasible alternatives for the site of a dam within the water course of a large river, selection of a site, and survey for design of a dam to be constructed on the selected site.

2. Land Use—Grazing, recreation, and wild life protection.

3. Topography—Mountainous.

4. Characterization—Level two for general configurations of the topography for site determination and selection, and level one for design of the dam.

5. Map-Compilation Factor—One or two, whichever is photogrammetrically feasible.

The criteria establish this as an engineering project requiring a survey of most of the river course in a mountainous area to determine site alternatives, to select one from the many, and to survey the selected site for design and construction of a dam thereon to impound water upstream therefrom. The constructional suitability of each feasible site and the acre feet of water to be stored upon construction of the dam within the confines of the river course and its tributaries will influence survey of area and alternative site mapping, and site selection.

Using the meter instead of the foot as the measurement unit, steps in selection of topographic maps for acquisition from available sources or for their design and compilation, as necessary, for use in accomplishing the foregoing engineering work for this project are:

1. Refer to Table 1. The first functional use of topographic maps and augmenting maps, as needed, is Survey of the water storage Area throughout the river and its tributaries from its stream bed to an elevation higher than the elevation to which the top of the proposed dam will be constructed after each feasible site for it has been identified, compared, and one selected. Thus, the first functional use of maps for this project has the E classification.

2. Use Table 2. Select from opposite Reservoir, within the Hydraulic and Hydrologic Development classification, from Column E the Map Group Number 13, because the characterization level is two for general configurations of the topography.

3. Use Table 3-M. Opposite Map Group Number 13, Column 1, select the contour interval range of 2.5 to 40 meters and the map scale Number range of 25,000 to 40,000 for Rangeland, Column 6.

4. Use Table 4-M. For map use E, Column 1, select opposite Map Group Number 13, Column 2, the map scale of 1:25,000, Column 3, and the contour interval of 15 meters, which is the median, Column 17, for mountainous topography. Do this because the characterization level is two for this stage of the engineering project.

The contour-interval-to-map-scale ratio, as a truncated value, is 1/1,600 for this topographic mapping. Thus, it is possible to use the map-compilation factor of one, with assurance that specification accuracies can be maintained easily throughout compilation of the maps by aerial photogrammetric methods, if the ground is not too obscured by tall and dense vegetation.

5. Refer to Table 1. The second functional use of topographic maps for this project has the D classificaiton, inasmuch as the characterization level is two for comparing the alternative sites and selecting one from the many for design and construction of the dam.

6. Refer to Table 2. Opposite Dams, within the Hydraulic and Hydrologic Development classification, select from Column D Map Group Number 11 within the range of 10 to 12.

7. Use Table 3-M. Opposite Map Group Number 11, Column 1, select the contour interval range of 1.25 to 20 meters and the map scale number range of 8,000 to 10,000 for Rangeland, Column 6.

8. Use Table 4-M. For functional map use D, Column 1, select opposite Map Group Number 11, Column 2, the map scale Number 8,000, Column 3, and the contour interval of 8 meters, which is the median for mountainous topography, Column 17.

The contour-interval-to-map-scale ratio of 1/1,000 within the maps of each alternative site for the dam is so large that there would be only two limitations on use of photogrammetric methods of map compilation, and these are:

(a) Whether or not the aircraft could be flown at a small enough flight height, while attaining adequate safety within the mountainous area, for taking aerial photographs at a scale large enough to attain the advantage of that ratio for each site, and

(b) If the ground within the bounds of each alternative site for the dam is not obscured too much by tall and dense vegetation.

9. Refer to Table 1. The third functional use of maps for this engineering project has the B classification for General Design of the dam at the site selected.

10. Use Table 2. Select from opposite Dams within the Hydraulic and Hydrologic Development classification from Column B the Map Group Number 7, because the characterization level is one for design of the dam and the topography is rugged.

11. Use Table 3-M. Opposite Map Group Number 7, Column 1, select the contour interval range of 0.3 to 4 meters and the map scale of 1:1,000 for Rangeland, Column 6.

12. Use Table 4-M. For Map use B, Column 1, select opposite Map Group Number 7, Column 2, and the map scale of 1:1,000, Column 3, the contour interval of 1.25 meter, which is the minimum for mountainous topography, Column 16. Do this because the characterization level is one for design of the dam.

The contour-interval-to-map-scale ratio of 1/800 of the maps required for the selected site of the dam is large enough to admit use of photogrammetric methods for map compilation. The restrictions to use of such methods would be (1) the flight height required for (a) safety reasons, (b) obtaining adequate endlap within each flight strip of vertical photographs, and (c) obtaining sufficient sidelap of parallel strips of the vertical photographs; and (2) vegetation, if too tall and dense at the selected site.

If the ground slopes are too steep for delineation of contours on the maps at the interval of 1.25 meter, index contours [every fourth contour for such an interval] should be used by omission of the intermediate three contours. Thus, within such topographic areas, the contour interval would become 5 meters. As an example, if the ground slope is 1:1 the distance between the index contours would be 0.5 of a centimeter at the compilation scale of 1:1,000, and 0.25 centimeter at the same scale if the ground slope is 0.5:1.

CONCLUSION

The preceding five examples are not all-inclusive. They were presented to serve as a guide to each user of the data provided herein. The criteria presented comprise a definition of and guidance by basic facts for each project, facility, or works. Use of the charts and tables, according to the pattern of selection indicated within the examples, will lead to determining efficiently and effectively the contour interval and scale of each map that will fulfill the needs of engineering, planning, and associated work. This will be so regardless of the character of the project, facility, or works, or of the type of topography, ground structure, soils, drainage, climate, vegetation, land uses, and so forth within the area(s) of concern.

Chapter 4.   ENGINEERING MAP ACCURACY STANDARDS AND TESTING

By: Dean C. Merchant, M., ASCE[1]

INTRODUCTION

Maps for engineering purposes must have accuracy clearly and unambiguously defined in terms familiar to the ultimate users of the maps. Producing maps possessing greater accuracy than required is wasteful; producing maps with less accuracy than required can be disastrous. Accordingly, this chapter provides definitions of accuracy at full scale[2] and in terms generally accepted by the engineering community. Specifications for field test procedures to assure that the required map accuracies have been attained are also presented.

Criteria for the selection of a map or maps to serve a specific purpose comprise the prudent and effective use of facts pertaining to contour interval, content by planimetric details, scale, and accuracy. The coordinate system and the plane of projection used in compiling the map(s) are also important. Each of the foregoing is affected by the functional use limitations inherent in maps. Such limitations are influenced by the contour-interval-to-map-scale ratio, although the contour interval and scale are appropriately correlated with character of the topography, type and intensity of land use, drainage detail, and actual plotting and delineation limitations.

Prior to World War II, maps were produced in the United States at many scales, by a variety of methods, and with widely different accuracies. Concurrently, the U.S. Geological Survey was engaged in the preparation of a standard topographic map series intended eventually to cover the entire United States. In response to the obvious need to standardize the map compilation and production of the many government and private agencies to permit possible inclusion in the national map atlas, a series of attempts was made to establish national map accuracy standards. The history of these efforts has been well-documented in the paper by L.E. Marsden [1960] of the U.S. Geological Survey. Continuing efforts to establish uniform standards ultimately resulted in the adoption of the United States National Map Accuracy Standards by the Bureau of the Budget in June 1947. These standards are applicable for all scales of general area coverage maps and have served for the last thirty years for small-scale maps.

With the introduction of aerial photogrammetric methods for the

---

[1] Professor, Department of Geodetic Science and Surveying, The Ohio State University, Columbus, Ohio.

[2] Refers to the mapped object itself

towards engineering and associated uses, can be characterized as diversified in contour interval, scale, accuracy, and content. The National Map Accuracy Standards, although providing specifications for maps at scales larger than 1:20,000, as well as for smaller scales, do not lend themselves readily to these varied requirements for maps in diversified categories.

In an effort to satisfy the need for greater flexibility in producing maps for engineering and associated purposes, at the same time preserving an unambiguous standard of accuracy, and with an added purpose of using more generally understood error terminology at the full scale of the mapped feature or detail (permitting more effective communication between the engineer and the mapper), an alternative to the U.S. National Map Accuracy Standards for large-scale maps is proposed. The alternative standard, entitled, "Engineering Map Accuracy Standards," is subsequently presented.

ENGINEERING MAP ACCURACY STANDARDS

A high degree of flexibility is essential for satisfying the diversified requirements in the detail and accuracy of maps used for engineering and associated purposes. Accordingly, these accuracy standards are offered as an alternative to that portion of the U.S. National Map Accuracy Standards (USNMAS) which pertain to large-scale maps (Refer to Appendix A). By USNMAS definition, any map is large in scale which is compiled and published or used at a scale larger than 1:20,000. The accuracy alternatives included herein are outlined in principle. The specific numerical value applied to each of the types of errors considered and presented should be governed by use requirements, feasibility, and cost of compilation.

ENGINEERING MAP ACCURACY STANDARDS

1. For engineering maps at any scale, Engineering Map Accuracy Standards (EMAS) are defined in terms of allowable or limiting errors.

   (a) The limiting standard errors[1] (limiting "Standard Deviation" (σ_0) [Burington, May 1970]) are estimated from position discrepancies determined at twenty or more well-defined and widely distributed points. These discrepancies are determined by comparison of the mapped position converted to full scale to the position of the corresponding points on the ground (object) as determined independently by means of a survey of adequate accuracy.[2] These points will have been withheld during the mapping procedure. The limiting standard errors shall not exceed:

   _____meter(s) [or feet] in X (plan)
   _____meter(s) [or feet] in Y (plan)
   _____meter(s) [or feet] in Z (elevation)

   (b) The limiting mean absolute error[3] (limiting absolute value of the "mean" deviation ($|\bar{\delta}_0|$) [Burington, May 1970]) is estimated as the absolute value of the mean position discrepancy at twenty or more well-defined and widely distributed horizontal/vertical points. These positions are determined from the map and independently by means of a survey of adequate accuracy. The positions of these points will have been withheld during the mapping procedure. The value of ($|\bar{\delta}_c|$) shall not exceed:[4]

   _____meter(s) [or feet] in X (plan)
   _____meter(s) [or feet] in Y (plan)
   _____meter(s) [or feet] in Z (elevation)

   (c) Compliance Testing:
   The map is tested to assess compliance with the specified limiting errors. A test for bias is performed using the limiting mean absolute error $|\bar{\delta}_0|$ and hypothesis testing based on the 't-distribution' within a 95% confidence interval ($1-\alpha$) and a one tailed test. A test for precision is performed using the limiting standard errors (σ_0) and hypothesis testing based on the 'Chi^2 (χ^2) distribution' within a 95% confidence interval ($1-\alpha$) and a one tailed test.

Discrepancies exceeding three times the specified limiting
standard errors will be construed as blunders and will be
corrected regardless of the results of other accuracy compliance tests.

2. The selection of well-defined test points is to be accomplished
with due consideration for both the character of the image presented to the map compiler and the deliberate shift of the
position of features to accommodate standard map symbolization.
Points selected for testing vertical accuracy compliance need
not be the same as those selected for horizontal compliance
testing. Points should be chosen in widely distributed locations
which are representative of the total map or map subregion.
Location of test points will be such that they are separated
a minimum of 1/12 and a maximum of 1/4 the diagonal dimension
of the map coverage. At least 15% of the test points shall appear in each quadrant of the map. In regions of sparse detail,
a target of suitable shape and color contrast shall be set on
the ground in the vicinity of each test point, and, in turn,
its horizontal position shall be plotted on the map to assure
unique identification of the test point on the ground.

3. Engineering maps meeting Engineering Map Accuracy Standards
will be so identified by means of a note appearing in the
map legend as follows:

This map complies with the Engineering Map Accuracy
Standards at a scale of _____ with error limits
not exceeding:

meter(s) [or feet]			
error type	X	Y	Z
σ_0			
$|\delta_0|$			

4. A particular map of uniform scale may contain regions, or
areas, where accuracy standards may differ greatly from the
standards applicable throughout the other portions of the
map. For such regions, or areas, an appropriate statement
should be included within the map legend. The region, or
areas, to which it applies should be clearly delineated and
identified on a reduced scale diagram of the map.

5. These standards are intended to provide a fundamental standard of map accuracy to facilitate communication between the
users of specialized maps for engineering and associated
purposes and those who compile them. For the well-developed
applications, such as maps for highway location, and design,

and right-of-way evaluation and procurement, a more detailed
standard may exist [American Society of Photogrammetry 1968].

(1) Limiting standard error (σ_0) is estimated (s) for each coordinate axis from:

$$s_X = \left\{ \left[\sum_{i=1}^{n} (\delta X_i - \bar{\delta}X)^2 \right] /(n - 1) \right\}^{\frac{1}{2}}$$

$$s_y = \left\{ \left[\sum_{i=1}^{n} (\delta Y_i - \bar{\delta}Y)^2 \right] /(n - 1) \right\}^{\frac{1}{2}}$$

$$s_z = \left\{ \left[\sum_{i=1}^{n} (\delta Z_i - \bar{\delta}Z)^2 \right] /(n - 1) \right\}^{\frac{1}{2}}$$

where:

δ_i = the discrepancy between the full scale position of a
well-defined feature determined from map measurements and
the position determined by means of the independently surveyed position for the point (i) in any given coordinate
direction.

$\bar{\delta}$ = the mean algebraic deviation in any given coordinate
direction.

$$\bar{\delta} = \left\{ \sum_{i=1}^{n} \delta_1 \right\} /n$$

n = the number of observed discrepancies (here taken as $\geq$ 20).

(2) The term "adequate accuracy" denotes a check survey accuracy
which is estimated to be at least equal to that of the control
survey from which the map is compiled. The check survey may
be conducted and adjusted along with the survey for map compilation purposes but the coordinate values of the check points
must be withheld during compilation.

In no case shall the check survey design error (e) exceed
one-third (1/3) of either the specified limiting standard
error (σ_0) or limiting mean absolute error ($|\bar{\delta}_0|$). The
order and class of survey is selected from and conforms to
the standards of accuracy and specifications published by

NOAA of the U.S. Department of Commerce. The diagonal full scale distance of map coverage (K) and the check survey error (e) is used to compute the required check survey proportional accuracy. The order and class of required horizontal check survey is determined from Table 4, of "Classification, Standards of Accuracy, and General Specifications of Geodetic Control Surveys" [NOAA, 1974] corresponding to the entry titled "Nominal accuracy or precision between adjacent points". Similar check survey selection procedures may be used for analytical photogrammetric methods provided similar survey specifications have been adopted by NOAA. The vertical check survey conforms to the same standards [NOAA, 1974] according to third-order methods corresponding to the entry titled "Nominal accuracy between points" where the term K is taken to mean the diagonal distance of map coverage.

(3) Limiting mean absolute error ($|\bar{\delta}_0|$) is estimated for each coordinate axis from

$$|\bar{\delta X}| = |\left\{ \sum_{i=1}^{n} (\delta X_i) \right\} /n|$$

$$|\bar{\delta Y}| = |\left\{ \sum_{i=1}^{n} (\delta Y_i) \right\} /n|$$

$$|\bar{\delta Z}| = |\left\{ \sum_{i=1}^{n} (\delta Z_i) \right\} /n|$$

where terms are defined in footnote (1).

EXPLANATION

   An explanation of the relationships between the Engineering Map
Accuracy Standards (EMAS) and the U.S. National Map Accuracy Standards
(USNMAS) is offered to facilitate interpretation. Most relationships
between error types assume some knowledge of the distribution of the
errors or discrepancies. It is most generally assumed that they are
normally distributed when in practice the discrepancies may well be
subject to some additional underlying systematic error. Analysis of
errors under such usual circumstances, although quite possible in
theory, require such sophistication that the analysis becomes
impractical. With some compromise of the theory, the process of error
analysis can become a realistic practice. The compromise is the
acceptance of the assumption that after the "mean absolute error "(an
average bias systematic error) has been removed from the discrepancies,
the resulting altered discrepancies are normally distributed. If this
assumption is acceptable, the errors (measures of precision), as usually
defined, can be estimated realistically and direct analysis applied in
practice according to the accepted definitions. The EMAS are based on
the acceptance of this compromise.

   The EMAS were developed on the premise that the limiting errors are
estimated from discrepancies (δ_i) between the full-scaled values
determined for a mapped feature's position obtained through measurements
on the map and values determined independently by a survey of adequate
accuracy. The comparison checks are made on a point-by-point basis at
the full scale of the feature or mapped detail. Ill-defined features,
or those subject to cartographic generalization, are not assumed to
conform with the specified accuracy and are omitted from any accuracy
evaluations under the EMAS.

   The 'mean absolute error' ($|\bar{\delta}|$) is defined as the absolute value of
the algebraic mean of the discrepancies in each of the coordinate
directions. [See footnote (3) of the EMAS]. The 'mean absolute error'
in any coordinate direction is estimated as the average discrepancy in
that direction measured at the check points. The algebraic signs
(senses) of the discrepancies are retained but the algebraic sign
(sense) of the final absolute average is not. The 'limiting standard
error' (σ_0) is estimated as the standard deviation from the mean
deviation. [See footnote (1) of the EMAS]. That is, the estimate of
the 'limiting standard error' (s) in any coordinate direction is the
square root of the sum of the squared discrepancies at the check points
in any coordinate direction after altering to account for the average of
the discrepancies at the check points and after dividing the sum of the
squared discrepancies by the number of check points less one. An
example of the computation of these two error quantity estimates is
given in Appendix VI.

   The horizontal components of the limiting standard errors may be
related to the conventional map accuracy statements of the USNMAS, which
have been termed by others as the "Circular Map Accuracy Standards"
(CMAS). The positioning error of features (details) according to the
CMAS is the radius of a horizontal circle within which 90 percent of all
discrepancies of well-defined points have been plotted. The CMAS may be

related to the horizontal component standard deviation (σ_x, σ_y) with sufficient approximation by [ACIC, 1962, pp. 59][1];

$$CMAS = 1.073 \ (\sigma_x + \sigma_y),$$

provided

$$\sigma min/\sigma max \geq 0.2$$

It is assumed that $\sigma_x = \sigma_y$, a reasonable assumption in a well-balanced photogrammetric solution, the relationship between the horizontal components of standard errors (EMAS System) and the CMAS error definitions (USNMAS) becomes:

$$\sigma_x = \sigma_y = 0.466 \ CMAS$$

A similar relationship can be developed for accuracy definitions in elevation. The USNMAS in elevation states that no more than 10 percent of all elevations tested (as determined from contours on the map) shall be in error in excess of one-half a contour interval. This allowable error may be termed Vertical Map Accuracy Standard (VMAS). The allowable error in elevation determined from the contours may be related to the allowable standard error in elevation by reference (1):

$$\sigma_z = 0.608 \ VMAS$$

In summary:

	USNMAS	Multiplied by	To Obtain EMAS
Horizontal	CMAS	0.466	$= \sigma_x$ or σ_y
Vertical	VMAS	0.608	$= \sigma_z$

Note that in regions of steeply sloping terrain, it is the practice to omit from the map contours that are intermediate between the indexing (elevation labled) contours. Since the vertical component of the USNMAS is based on the plotted contour interval, the EMAS vertical component is also increased in magnitude in regions (areas) where the intermediate contours are omitted from the map and only the index contours are plotted.

Wherever maps for engineering and associated purposes need not be related through a common coordinated system to other maps or positional information, the (mean absolute error) ($|\bar{\delta}_0|$) would be undefined. The result would be a map which is internally consistent in accordance with the specified standard error but would bear no relationship to any other coordinate system. Phrased in another way, this would be a map based on a local coordinate system.

In the event that a local coordinate system is adopted for map control (i.e., $|\bar{\delta}_0|$ is unspecified), the 'survey of adequate accuracy' is itself a local survey, internally accurate, but independent of any other system.

TYPICAL EXAMPLES

Examples are offered in Table 1 in which the well-known error definitions provided by the USNMAS are related to corresponding EMAS values in terms of standard errors along Cartesian coordinate axes. Examples are given in both the SI (Systeme International des Unites) system and the English foot system of measure.

Figure 1 indicates graphically the fixed relationships which prevail between the U.S. National Map Accuracy Standards (USNMAS) and the alternative for large scales, the Engineering Map Accuracy Standards (EMAS). Both the forward and back examples of the transformation given earlier can be adequately and readily interpreted from this graph. The graphical relationships of Figure 1 are based on the assumption that systematic errors are negligible. The USNMAS makes no reference to the systematic components of map errors. Consequently, there is no corresponding term in the USNMAS to that of 'limiting mean absolute error', as defined in the EMAS.

71

TABLE 1. Examples of Transformation Between Map Characteristics
in the USNMAS and the EMAS

SI SYSTEM	ENGLISH SYSTEM
Example No. 1	Example No. 3

Example No. 1

  Given: USNMAS

    scale = 1:10,000

    CI = 1 meter

  Then in plan:

    CMAS = 1/30 (0.0254) 10,000

        = 8.5 meters

$\therefore \sigma_x = \sigma_y$ = CMAS (0.466)

        = 4.0 meters

In elevation:

    VMAS = ½CI = 0.5 meter

    $\therefore \sigma_z$ = VMAS (0.608)

        = 0.3 meter

Example No. 2

  Given: EMAS

    σ_x = 1.0 meter

    σ_y = 1.0 meter

    σ_z = 0.1 meter

  Then, in plan:

    provided $\sigma_x = \sigma_y$

    CMAS = σ_x/0.466

        = 2.1 meter

In elevation:

    CI = $2\sigma_z$/0.608 = 0.33 meter

Corresponding to Scale No. for
  USNMAS

  = CMAS $(1/30 (0.0254))^{-1}$ = 2,500

for EMAS

    not applicable

Example No. 3

  Given: USNMAS

    scale = 1:2,400

    CI = 1 foot

  Then in plan:

    CMAS = 1/30 (1/12) 2,400

        = 6.7 feet

$\therefore \sigma_x = \sigma_y$ = CMAS (0.466)

        = 3.1 feet

In elevation:

    VMAS ½CI = 0.5 feet

    $\therefore \sigma_z$ = VMAS (0.608)

        = 0.3 feet

Example No. 4

  Given: EMAS

    σ_x = 3.0 feet

    σ_y = 3.0 feet

    σ_z = 0.3 feet

  Then, in plan:

    provided $\sigma_x = \sigma_y$

    CMAS = σ_x/0.466

        = 6.4 feet

In elevation:

    CI = $2\sigma_z$/0.608 = 0.99 feet

Corresponding to Scale No. for
  USNMAS

  = CMAS $(1/30 \cdot 1/12)^{-1}$ = 2,300

for EMAS

    not applicable

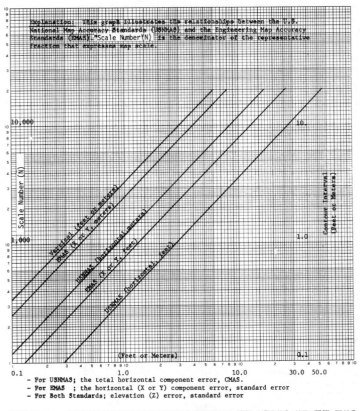

0.1   1.0   10.0   30.0 50.0

- For USNMAS; the total horizontal component error, CMAS.
- For EMAS ; the horizontal (X or Y) component error, standard error
- For Both Standards; elevation (Z) error, standard error

FIGURE 1. CORRESPONDING MAP ERRORS BETWEEN THE USNMAS AND THE EMAS
AS A FUNCTION OF SCALE NUMBER AND CONTOUR INTERVAL

73

DISCUSSION

There are functional use limitations which are inherent and must be considered when defining accuracy. Among these limitations are the essentiality of achieving consistency in the contour-interval-to-map-scale-ratio, according to need and to the use of the map. This concern is discussed in Chapters 2 and 3.

The contour interval and map scale, of course, must be correlated with the character of topography, the type and intensity of land use, the drainage details, and the plotting and delineation limitations. As an example, if the map scale were to be 1:1,200 (100 feet per inch) and the slopes of the ground were generally 1:1 and steeper, a contour interval of 1 foot (0.3m) would be too small to be practical. By contrast, if the map scale were 1:1,200 and the ground slopes were generally less steep than 20:1, a contour interval of 10 feet (3 m) would be unrealistically large.

For map standards for ordinary engineering purposes in which acceptable contour intervals, scales, and accuracy standards have been established through experience, reference should be made to Chapters 2 and 3. In some instances, the usual (USNMAS) will apply (i.e., $\frac{1}{2}$ contour interval, and 1/30 inch, (0.85 mm), for 90 percent of well-defined points; see Appendix I). However, for unusual or highly specialized purposes in which the engineer has unique requirements for accuracy, the Engineering Map Accuracy Standards (EMAS) are offered as a more definitive alternative for specifying the accuracy of full-scale position and dimension of the mapped feature or detail as determined from the map at its use scale. (See Appendix II)

Care must be taken in the application of the EMAS and its associated CMAS and VMAS to assure that the limits of the photogrammetric plotting equipment are not exceeded. Allowances may be made for use of innovative photogrammetric compilation procedures through use of the 'Map-Compilation Factor' as discussed in Chapter 3. However, departures from current practice should be undertaken only after consultation with appropriate experts. Plotters based on analog principles are limited in the ratio of range of object distances to object depth that can be accommodated. Consideration should also be given to the degree of slope of the features being mapped. This will bear on the selection of the scale of the map to assure that the error in the plotted position of contours does not introduce excessive error into the representation of elevation.

Contours and planimetric details on topographic maps compiled photogrammetrically cannot have the same point-by-point accuracy as measurements made of finite points on an individual basis, or on a repetitive and averaging basis, as is done for testing or for experimental purposes. This is because the contours and the planimetric details represent continuous positioning of the measuring mark by the operator of the photogrammetric instrument or system as the mark is moved over the surface of the ground, or along the edge of the feature or detail being delineated planimetrically, as each is viewed within the stereoscopic model. Moreover, the measuring mark, as it is moved, is not always exactly on the surface of the ground for contours, or on the edge of other essential details, as the mapping is being done. Such facts emphasize realistically the reason for variations in accuracy of both contours and planimetry within maps compiled photogrammetrically, and they also make necessary definitions on a practical basis of the limit in errors which can be tolerated. The accuracy standards and testing procedures presented herein fulfill that need.

An example of accuracy compliance testing of the map is presented in Appendix VI. The testing procedure is based on comparisons of full-scale coordinates of well-defined features as determined from the map to those of corresponding features determined by a check survey. The design of the check survey requires a comprehensive knowledge of the standards and specifications published by the National Oceanic Atmospheric Administration. It is advisable to acquire the services of a consultant knowledgeable in control surveys if this guidance is not otherwise available.

SUMMARY

The United States National Map Accuracy Standard (USNMAS) has served since 1941 for maps at scales smaller than 1:20,000. Due to the broad range of special applications of large-scale maps, particularly for engineering purposes, an alternative accuracy standard termed the Engineering Map Accuracy Standard (EMAS) is presented. It is intended that the new EMAS will facilitate understanding of map accuracy as required by the map users and as interpreted by the map producers.

APPENDIX I - UNITED STATES NATIONAL MAP ACCURACY STANDARDS, U.S. BUREAU OF THE BUDGET; Issued June 10, 1941; Revised April 26, 1943; Revised June 17, 1947

With a view to the utmost economy and expedition in producing maps which fulfill not only the broad needs for standard or principal maps, but also the reasonable particular needs of individual agencies, standards of accuracy for published maps are defined as follows:

1. Horizontal accuracy. For maps on publication scales larger than 1:20,000, not more than 10 percent of the points tested shall be in error by more than 1/30 inch, measured on the publication scale; for maps on publication scales of 1:20,000 or smaller, 1/50 inch. These limits of accuracy shall apply in all cases to position of well-defined points only. 'Well-defined' points are those that are easily visible or recoverable on the ground, such as the following: monuments or markers, such as bench marks, property boundary monuments; intersections of roads, railroads, etc.; corners of large building or structures (or center points of small buildings); etc. In general what is 'well-defined' will also be determined by what is plottable on the scale of the map within 1/100 inch. Thus, while the intersection of two roads or property lines meeting at right angles would come within a sensible interpretation, identification of the intersection of such lines meeting at an acute angle would obviously not be practicable within 1/100 inch. Similarly, features not identifiable upon the ground within close limits are not to be considered as test points within the limits quoted, even though their positions may be scaled closely upon the map. In this class would come timber lines, soil boundaries, etc., etc.

2. Vertical accuracy, as applied to contours on the maps at all publication scales, shall be such that not more than 10 percent of the elevations tested shall be in error more than one-half the contour interval. In checking elevations taken from the map, the apparent vertical error may be decreased by assuming a horizontal displacement within the permissible horizontal error for a map of that scale.

3. The accuracy of any map may be tested by comparing the positions of points whose locations or elevations are shown upon it with corresponding positions as determined by surveys of a higher accuracy. Tests shall be made by the producing agency, which shall also determine which of its maps are to be tested, and the extent of such testing.

4. Published maps meeting these accuracy requirements shall note this fact in their legends, as follows: "This map complies with the National Map Accuracy Standards."

5. Published maps whose errors exceed those aforestated shall omit from their legends all mention of standard accuracy.

6. When a published map is a considerable enlargement of map drawing ('manuscript') or of a published map, that fact shall be stated in the legend. For example, "This map is an enlargement of a 1:20,000

scale map drawing," or "This map is an enlargement of a 1:24,000 scale published map."

7. To facilitate ready interchange and use of basic information for map construction among all federal mapmaking agencies, manuscript maps and published maps, wherever economically feasible and consistent with the uses to which the map is to be put, shall conform to latitude and longitude boundaries, being 15 minutes of latitude and longitude, or 7 1/2 minutes, or 3 3/4 minutes in size.

APPENDIX II. - SELECTION OF APPROPRIATE ERROR VALUES

Specific error values are not suggested in the Engineering Map Accuracy Standards (EMAS) due to the variability of character and application for this map type. This is in contrast to the United States National Map Accuracy Standards (USNMAS) in which the goal was to standardize accuracies, contour intervals, and scales to permit maps to be adopted into the established National Atlas. For large-scale maps, the process of selection of coordinate system definition, contour interval, and values of accuracies in terms of limiting errors (standard error) is left to negotiations between the engineering client and the cartographic engineer. A comprehensive treatment of the process for selection of appropriate map scales in accord with the intended map use is presented in Chapter 3. Although it is not the function of the EMAS itself to specify error values, the completed specification requires specific values of limiting standard errors along each axis. These are in terms of ground values and are the only values used for testing the compliance of the map with the specified accuracy according to the EMAS.

It should be recognized that practical limitations exist on the accuracy which can be feasibly attained at reasonable cost. Also, the money which should be expended to attain a particular accuracy is influenced by the magnitude of the project area, the extent and density of existing control, and the character of the region in which the control surveys are to be made.

Given the choice of map scale guided by Chapter 3. and given the choice of coordinate accuracies in terms of limiting standard errors as selected by the engineering client for his specific application, it is useful to relate these choices to their corresponding values for CMAS and to the "maximum" error. Table B-1 presents these relationships between typical values of map scale, limiting standard errors and the resulting CMAS. Both the metric and English units are presented. The maximum error in the EMAS is defined as three times the limiting standard error. Three values of maximum and limiting standard error and the resulting CMAS are given for each map scale. The table presents data representing a practical range of CMAS values. The largest value is about 1/30 (0.33) inch or 0.85 mm; the smallest is about 1/50 (0.20) inch or 0.51 mm. The intermediate value of CMAS, about 1/40 (0.25) inch or 0.64 mm, is recommended for most engineering applications. The Circular Map Accuracy Standard (CMAS), discussed in Chapter 4 under "Explanation", is the radius of a circular error of 90 percent probability at map scale.

For those engaged in large-scale topographic mapping by photogrammetric methods, there is no substitute for the experience with the system employed when estimating accuracy obtainable in terms of the full-scale mapped feature. (See section of Chapter 4 headed "Discussion").

No limitation on choices of scale and limiting standard error are imposed by this accuracy standard. Such choices are discussed in Chapters 2 and 3. For those cases which fall between or beyond the entries in Table B-1, the following equations serve to compute the corresponding CMAS.

For the metric case, given that the units of the CMAS are those of the limiting standard error (σ) and that scale is in terms of the representative fraction then: (metric) CMAS = (σ . scale)/0.466

For the English case, given that the units of the CMAS are in inches, the limiting standard error (σ) is in feet and the scale is in terms of "feet per inch", then: (English) CMAS = σ/ (scale . 0.4666)

For elevation accuracy in terms of the limiting standard error (σ), the relationship between (σ) and contour interval (CI) according to the EMAS for either metric or English systems is always:

$$\sigma = 0.304 \cdot CI$$

TABLE B−1

** MAP ACCURACY TABLE **

METRIC ********** ENGLISH

Scale Number N	X or Y (meters) Max.	X or Y (meters) Standard	CMAS (mm)	:::	Scale feet/ inch	X or Y (feet) Max.	X or Y (feet) Standard	CMAS (inch)
50	0.06	0.02	0.86		2	0.09	0.03	0.032
50	0.03	0.01	0.43		2	0.06	0.02	0.021
50	0.03	0.01	0.43		2	0.06	0.02	0.021
60	0.06	0.02	0.72		4	0.18	0.06	0.032
60	0.06	0.02	0.72		4	0.15	0.05	0.027
60	0.03	0.01	0.36		4	0.12	0.04	0.021
100	0.12	0.04	0.86		5	0.24	0.08	0.034
100	0.09	0.03	0.64		5	0.18	0.06	0.026
100	0.06	0.02	0.43		5	0.15	0.05	0.021
125	0.15	0.05	0.86		8	0.36	0.12	0.032
125	0.12	0.04	0.69		8	0.27	0.09	0.024
125	0.09	0.03	0.52		8	0.21	0.07	0.019
250	0.30	0.10	0.86		10	0.48	0.16	0.034
250	0.21	0.07	0.60		10	0.36	0.12	0.026
250	0.18	0.06	0.52		10	0.27	0.09	0.019
400	0.48	0.16	0.86		20	0.93	0.31	0.033
400	0.36	0.12	0.64		20	0.69	0.23	0.025
400	0.27	0.09	0.48		20	0.57	0.19	0.020
500	0.60	0.20	0.86		30	1.41	0.47	0.034
500	0.45	0.15	0.64		30	1.05	0.35	0.025
500	0.36	0.12	0.52		30	0.84	0.28	0.020
600	0.72	0.24	0.86		40	1.86	0.62	0.033
600	0.54	0.18	0.64		40	1.41	0.47	0.025
600	0.42	0.14	0.50		40	1.11	0.37	0.020
800	0.96	0.32	0.86		50	2.34	0.78	0.033

Table 1 continued

800	0.72	0.24	0.64	50	1.74	0.58	0.025
800	0.57	0.19	0.51	50	1.41	0.47	0.020
1000	1.17	0.39	0.84	60	2.79	0.93	0.033
1000	0.90	0.30	0.64	60	2.10	0.70	0.025
1000	0.72	0.24	0.52	60	1.68	0.56	0.020
1250	1.50	0.50	0.86	80	3.60	1.20	0.032
1250	1.11	0.37	0.64	80	2.79	0.93	0.025
1250	0.90	0.30	0.52	80	2.25	0.75	0.020
1500	1.80	0.60	0.86	100	4.80	1.60	0.034
1500	1.20	0.40	0.57	100	3.60	1.20	0.026
1500	1.08	0.36	0.52	100	2.79	0.93	0.020
2000	2.40	0.80	0.86	160	7.50	2.50	0.034
2000	1.80	0.60	0.64	160	5.70	1.90	0.025
2000	1.50	0.50	0.54	160	4.50	1.50	0.020
2400	2.70	0.90	0.81	200	9.30	3.10	0.033
2400	2.10	0.70	0.63	200	6.90	2.30	0.025
2400	1.80	0.60	0.54	200	5.70	1.90	0.020
2500	3.00	1.00	0.86	250	11.70	3.90	0.033
2500	2.10	0.70	0.60	250	8.70	2.90	0.025
2500	1.80	0.60	0.52	250	6.90	2.30	0.020
3000	3.60	1.20	0.86	400	18.60	6.20	0.033
3000	2.70	0.90	0.64	400	14.10	4.70	0.025
3000	2.10	0.70	0.50	400	11.10	3.70	0.020
5000	6.00	2.00	0.86	500	23.40	7.80	0.033
5000	4.50	1.50	0.64	500	17.40	5.80	0.025
5000	3.60	1.20	0.52	500	14.10	4.70	0.020
6000	7.20	2.40	0.86	600	27.90	9.30	0.033
6000	5.40	1.80	0.64	600	21.00	7.00	0.025
6000	4.20	1.40	0.50	600	16.80	5.60	0.020
8000	9.60	3.20	0.86	800	36.00	12.00	0.032
8000	7.20	2.40	0.64	800	27.90	9.30	0.025
8000	5.70	1.90	0.51	800	22.50	7.50	0.020
10000	11.70	3.90	0.84	1000	48.00	16.00	0.034
10000	9.00	3.00	0.64	1000	36.00	12.00	0.026
10000	7.20	2.40	0.52	1000	27.90	9.30	0.020
12500	15.00	5.00	0.86	1600	75.00	25.00	0.034
12500	11.10	3.70	0.64	1600	57.00	19.00	0.025
12500	9.00	3.00	0.52	1600	45.00	15.00	0.020
15000	18.00	6.00	0.86	2000	93.00	31.00	0.033
15000	12.00	4.00	0.57	2000	69.00	23.00	0.025
15000	10.80	3.60	0.52	2000	57.00	19.00	0.020
20000	24.00	8.00	0.86				
20000	18.00	6.00	0.64				
20000	15.00	5.00	0.54				
24000	27.00	9.00	0.81				
24000	21.00	7.00	0.63				
24000	18.00	6.00	0.54				
25000	30.00	10.00	0.86				
25000	21.00	7.00	0.60				
25000	18.00	6.00	0.52				

APPENDIX III - REFERENCES

1. ACIC (1962) "Principles of Error Theory and Cartographic
   Applications" Aeronautical Chart and Information Center, Technical
   Report No. 96, February, 1962.

2. American Society of Photogrammetry, Photogrammetry for Highways
   Committee (1968) "Reference Guide Outline, Specifications for Aerial
   Surveys and Mapping by Photogrammetric Methods for Highways",
   published by U.S. Dept. of Transportation, Federal Highway
   Administration, 1968.

3. Burington, May (1979), "Handbook of Probability and Statistics with
   Tables", McGraw Hill, 1970.

4. Hamilton, Walter C. (1964) "Statistics in Physical Science -
   Estimation, Hypothesis Testing, and Least Squares", Ronald Press
   Co., 1964.

5. Marsden, Lloyd E. (1960), "How the National Map Accuracy Standards
   were Developed", Surveying and Mapping, ACSM, Volume xx, No. 4.,
   pages 429-439.

6. Task Committee for the Preparation of a Manual on Selection of Map
   Types, Scales, and Accuracies for Engineering and Planning (1972)
   "Selection of Maps for Engineering and Planning", Journal of the
   Surveying and Mapping Division, ASCE, July, 1972.

7. Thompson, Morris M. (1960). "A Current View of the National Map
   Accuracy Standards," Surveying and Mapping ACSM, Volume xx, No. 4,
   pages 449-457.

8. National Oceanic and Atmospheric Administration (1974)
   "Classification, Standards of Accuracy, and General Specifications
   of Geodetic Control Surveys" National Ocean Survey, U.S. Dept. of
   Commerce, Rockville, MD., Feb. 1974

APPENDIX IV - NOTATION

The following symbols are used in this paper:

CMAS = Circular map accuracy standard (horizontal)

$\overline{\delta}$ = Mean error

$\overline{\delta}_0$ = Limiting mean error

$|\overline{\delta}X_0|$, $|\overline{\delta}Y_0|$, $|\overline{\delta}Z_0|$ = Limiting mean absolute error in the respective survey coordinate axis directions

n = Number of observed discrepancies

σ_{X_0}, σ_{Y_0}, σ_{Z_0} = Limiting standard errors in the survey coordinate system X, Y, and Z axis directions, respectively.

s_X, s_Y, s_Z = Sample standard errors in the survey coordinate system X, Y, and Z axis directions, respectively.

$|\overline{\delta}X|$, $|\overline{\delta}Y|$, $|\overline{\delta}Z|$ = Sample mean absolute error in the respective survey coordinate axis directions.

VMAS = Vertical map accuracy standard

APPENDIX V - EXAMPLE OF EMAS SPECIFICATION

1-a. An example in SI units:

A topographic map is to be compiled according to the EMAS for a contour interval of 0.5 meter and a scale of 1:1,000. The mean absolute error (mae) is to be small enough to assure that the map detail will be positionally consistent with maps of contiguous areas. Specifications for this purpose might be:

a. The limiting standard error (σ_0) shall not exceed:

0.28 meter in X

0.28 meter in Y

0.15 meter in Z

b. The limiting mean absolute error ($\overline{\delta}_0$) shall not exceed:

0.15 meter in X

0.15 meter in Y

0.08 meter in Z

c. Discrepancies exceeding 0.84 meter in X, or Y, or 0.45 meter in Z shall be defined as blunders and shall be corrected regardless of the results of other accuracy compliance tests.

2-a. Accuracy notations in SI units:

This map complies with the Engineering Map Accuracy Standards at a scale 1:1,000 with error limits not exceeding:

Error Type	In Meters		
	X	Y	Z
Standard error (σ_0)	0.28	0.28	0.15
mean absolute error $\|\delta_0\|$	0.15	0.15	0.08

1-b An example in English units:

For a rolling to slightly mountainous area, a topographic map is to be compiled according to the EMAS for a contour interval of 5 feet at the scale of 100 feet per inch. The mean absolute error (mae) is to be small enough to assure that the map detail will be

83

positionally consistent with maps of contiguous areas, and the specifications for this purpose shall be:

    a. The limiting standard errors (σ_0) shall not exceed:

        1.17 foot in X

        1.17 foot in Y

        1.52 foot in Z

    b. The limiting mean absolute error ($\bar{\delta}_0$) shall not exceed:

        0.30 foot in X

        0.30 foot in Y

        0.40 foot in Z

    c. Discrepancies exceeding 3.5 feet in X or Y, or 4.5 feet in Z shall be defined as blunders and shall be corrected regardless of other accuracy tests.

2-b   Accuracy notations in English units:

The following accuracy notation will appear on the map legend: This map complies with the Engineering Map Accuracy Standards at a scale of 100 feet per inch, with error limits not exceeding:

Error Type	In Feet		
	X	Y	Z
Standard error (σ_0)	1.17	1.17	1.52
Mean absolute error $\|\bar{\delta}_0\|$	0.30	0.30	0.40

In specifying the accuracy for any topographic map, care must be taken to assure that the contour interval is consistent with the vertical error limits, and that the scale of the map is consistent with the horizontal error limits which shall not be exceeded, respectively.

APPENDIX VI - EXAMPLE OF COMPLIANCE TESTING

Testing of the map is accomplished by accepted statistical procedures on both the sample limiting mean absolute errors ($|\overline{\delta X}|$, $|\overline{\delta Y}|$, $|\overline{\delta Z}|$) to assess the presence of a significant bias and on the sample limiting standard errors (S_x, S_y, S_z) to assess compliance with precision requirements. Hypothesis testing is performed on sample means and sample standard errors independently on each of the planimetric coordinates (X) and (Y) and on elevation (Z).

1. Check Survey Design: - Refer to footnote #2 of the EMAS.

The check survey is designed such that the check survey error (e) will be less than 1/3 the magnitude of both the specified limiting standard error (σ_0) and the limiting mean absolute error ($\overline{\delta}_0$) along each coordinate axis independently. For this example see Appendix E. The check survey error (e) in X and Y survey coordinate directions is computed from the specified values of mean absolute error ($|\overline{\delta X_0}|$ or $|\overline{\delta Y_0}|$) as:

$$e_X = e_Y = \frac{|\overline{\delta X_0}|}{3} = \frac{0.2}{3} = 0.067 \text{ meters}$$

For the map scale of 1:1,000 in this example, the diagonal of full-scale ground coverage (D) could typically be 1,000 meters. (Accordingly, for the most stringent case, the value of $\frac{e}{D}$ corresponds to the "nominal accuracy or precision between adjacent points".) [NOAA, 1974, Table 4.] For this example, $\frac{e}{D} = 1/14,925$ Entering Table 4. [NOAA, 1974] it is seen that a Second-Order, Class II, of nominal accuracy of 1 part in 20,000 is the appropriate survey procedure for establishment of check points for this example.

The vertical check survey order and class is determined in a similar manner. The check survey error (e_Z) in elevation is computed from the specified mean absolute error ($|\overline{\delta Z_0}|$) in elevation as:

$$e_Z = \frac{|\overline{\delta Z_0}|}{3} = \frac{0.15}{3} = 0.05\text{m or } 50\text{mm.}$$

Entering Table 4. [NOAA, 1974] once more only for vertical surveys, it is clear that Third-Order leveling procedures will be more than adequate for the check survey. Note that in Table 4., the term (k) can be taken as the diagonal of the map at full-scale ground coverage (D) but in units of kilometers.

In summary, the requirement for testing for a bias of 0.20 meters in the X or Y coordinate directions requires a field survey procedure corresponding to Second-Order, Class II. [NOAA, 1974]. The test of elevation for bias of 0.15 meters requires only a Third-Order leveling procedure.

2. Testing for Bias:

The test for bias is made by testing the hypothesis at the 5 percent significance level (α). A one-tailed test for the hypothesis:

$$H_0: \mu = \mu_0$$

against the alternative

$$H_1: \mu > \mu_0$$

where: μ_0 is the specified limiting mean absolute error
μ is the expected limiting mean absolute error

For this, compute the statistic (t) for the (X) coordinate discrepancies:

$$t_X = \frac{(\,|\overline{\delta X}| - |\overline{\delta X_0}|\,)\ n^{\frac{1}{2}}}{s_X}$$

The hypothesis is accepted if:

$$|t_X| < t_{n-1, \alpha}$$

and the map is accepted as free from excessive bias error in the (X) coordinate direction. Tests for acceptance in the (Y) coordinate direction and in elevation (Z) are performed by the same procedure of hypothesis testing of univariate population. The statistic ($t_{n-1, \alpha}$) is drawn from a table of the "Student's t" distribution with one tail at a significance level of $\alpha = 5$ percent (confidence interval of 95 percent) and for (n-1) degrees of freedom. (see Table F2.)

3. Testing for Precision:

The test for precision is made by testing a hypothesis at the 5 percent significance level (α).

A one-tailed test for the hypothesis:

$$H_0: \sigma^2 = \sigma_0^2$$

against the alternative

86

$$H_1: \quad \sigma > \sigma_0^2$$

where: σ_0^2 is the specified standard error squared (specified variance)

σ^2 is the expected variance

For this, compute the statistic "Chi-squared" (χ^2):

$$\chi^2 = \frac{(n-1)\ s_x^2}{\sigma_0^2{}_x}$$

The hypothesis is accepted provided:

$$\chi^2 < \chi^2_{n-1,\alpha}$$

and the map is accepted as meeting the precision requirements in the (X) coordinate direction. Tests in the (Y) coordinate direction and in elevation (Z) are performed by the same procedure. The statistic ($\chi^2_{n-1,\alpha}$) is drawn from a table of the (χ^2) distribution of one tail at a significance level of $\alpha = 5$ percent and for (n-1) degrees of freedom. (see Table F2.)

The following fictitious example of computation of EMAS error values is based on use of 32 check points and is computed for the X-coordinate direction only. (See Table F-1). Similar computations would be performed for the (Y) and (Z) directions also. Assume the EMAS was specified as follows (see Appendix E.):

Error Type	Units of Length		
	X	Y	Z
Standard error (σ_0)	0.40	0.40	0.30
Mean absolute error $|\bar{\delta}_0|$	0.04	0.04	0.15

In this case, working with the data from Table F-1;
The number of test check points = n = 32

The mean algebraic deviation (sample mean) in the X–direction is

$$\overline{\delta X} = \left(\sum_{i=1}^{n} \delta X_i \right) / n$$

$$\overline{\delta X} = +0.13 \text{ and } |\overline{\delta X}| = 0.13$$

The estimate of limiting standard error (σ_0) = s_X

$$s_X = \left\{ \left[\sum_{i=1}^{n} (\delta X_i - \overline{\delta X})^2 \right] / (n-1) \right\}^{\frac{1}{2}} = 0.405$$

Testing for Bias:

$$t_X = \frac{(|\overline{\delta X}| - |\overline{\delta X_0}|) n^{\frac{1}{2}}}{s_X} = \frac{(|0.13| - |0.04|)(5.66)}{0.405} = 1.26$$

The map meets the specification for ($|\overline{\delta x}|$) provided:

$$|t_X| < t_{n-1,\alpha}$$

where ($t_{n-1,\alpha}$) is taken as a statistic based on the student's – t distribution using α = 5 percent, degrees of freedom = (n–1) and using a one–tailed assumption. For this case, the degrees of freedom are 31 and the corresponding t-statistic for 32 points taken from Table F2 is:

$$t_{n-1,\alpha} = 1.695$$

Accordingly, for these data, the map is accepted as meeting the specifications for mean absolute error.

Testing Precision:

$$\chi^2 = \frac{(n-1) s_X^2}{\sigma_{0_X}^2} = \frac{(31)(0.405)^2}{(0.4)^2}$$

$$\chi^2 = 31.8$$

The map meets the specifications for standard error (σ_{0X}) provided:

$$\chi^2 < \chi^2_{n-1,\,\alpha}$$

where ($\chi^2_{n-1,\alpha}$) is the statistic based on the chi-squared distribution using α = 5 percent, degrees of freedom = (n-1) and using a one-tailed assumption. For the case, the degrees of freedom remain 31 for 32 check points and the corresponding χ^2 statistic is:

$$\chi^2_{n-1,\,\alpha} = 44.97$$

Accordingly, for this example, the map is accepted as meeting the precision specification for standard error.

The test for existence of gross blunders requires that no discrepancy should exceed three times the limiting standard error.

$$3 \cdot \sigma_0 = 1.20$$

The discrepancies ($\delta X_i - \overline{\delta X}$) indicated in Table F -1 are all less than 1.20, consequently, the map is acceptable as free from gross blunders.

Tests for compliance in (Y) and in elevation (Z) are made in a similar manner for bias, precision, and existence of gross blunders.

Table F - 1

Example of Test Data
X-survey coordinate only

Point	δX_i	$\delta X_i - \overline{\delta X}$	$(\delta X_i - \overline{\delta X})^2$
1	.18	.05	.00
2	.12	-.01	.00
3	.00	-.13	.02
4	.61	.48	.23
5	.53	.40	.16
6	-.23	-.36	.13
7	.21	.08	.01
8	.24	.11	.01
9	-.29	-.42	.18
10	.21	.08	.01
11	-.32	-.45	.20
12	-.22	-.35	.12
13	-.08	-.21	.04
14	-.16	-.29	.08
15	.46	.33	.11
16	-.23	-.36	.13
17	.14	.01	.00
18	.55	.42	.18
19	-.24	-.37	.14
20	.21	.08	.01
21	-.23	-.36	.13
22	-.07	-.20	.04
23	.22	.09	.01
24	-.32	-.45	.20
25	1.22	1.09	1.19
26	-.20	-.33	.11
27	.23	.10	.01
28	.21	.08	.01
29	1.02	.89	.79
30	-.15	-.28	.08
31	.92	.79	.62
32	-.25	-.38	.14

TABLE F2. Statistics for Compliance Testing
(Significance Level $(\alpha) = 5$ %)

Numbers of check pts. (n)*	$t_{n-1,\alpha}$	$\chi^2_{n-1,\alpha}$
20	1.729	30.14
21	1.725	31.41
22	1.721	32.67
23	1.717	33.92
24	1.714	35.17
25	1.711	36.42
26	1.708	37.65
27	1.706	38.88
28	1.703	40.11
29	1.701	41.34
30	1.699	42.56
31	1.697	43.77
32	1.695	44.97

* Test statistics for the number of check points (n) > 32, can be drawn from other existing tables.

CHAPTER 5. MAP CONTENT AND SYMBOLS

By: Robert P. Jacober, Jr.[1]

INTRODUCTION

Maps answer the two questions "What?" and "Where?" (9). By doing so, maps become the communication link between the map designer and the map user. "The graphical content of a map may be presented as a symbolized delineation, as photographic imagery, or as a combination of symbolization and photographic imagery." (7). What maps portray, how they should depict what they portray, and the standardization of that depiction are the subjects discussed in this chapter.

Why standardize? In 1948, Walter Blucher, The Executive Director of the American Society of Planning Officials, made the following statement: "Here in the United States it is almost impossible to compare drawings prepared by different draftsmen or offices, not only because they may be of different scales, but because the symbols used are often as far apart as the poles." (18). This statement is as valid today as it was then. For example, the symbols in Fig. 1 are currently used by organizations to represent free standing light poles. If three of these symbols were selected to represent existing, proposed, and removed light poles, the confusion due to the multiplicity of symbols representing the same feature would be eliminated. The remaining symbols could then be used to depict other features.

Maps at the engineering scales are also very efficient storage devices. The amount of information that can be stored on one flat map would need a filing cabinet to store the data if verbal descriptions were used. But without standardized map symbols the data retrieval becomes inefficient or even incorrect. The purpose of this chapter is to suggest a set of map symbols and recommend procedures that, if adopted as a standard, could markedly improve intercommunication and data retrieval. In keeping with the scope of this special publication, certain limitations have been placed on the maps discussed and the symbols developed.

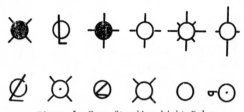

Figure 1   Free Standing Light Poles

---

[1] Major, U.S. Air Force, Cartographic Staff Officer, Defense
    Mapping Agency Aerospace Center, St. Louis, Missouri 63118.

Scale and purpose are the two main constraints on map content and symbology. This chapter focuses primarily on the content of maps between the scales of 20 feet per inch (1:240) and 400 feet per inch (1:4,800). Maps at these scales of 1:240 to 1:4,800 are used primarily to show the location of manmade features. The maps are divided into two classes, planning maps and operational or maintenance maps. Planning maps emphasize the symbols showing the existing features. Most of these maps are not usually mass produced. The production methods, whether by hand drafting, mechanical reproduction, or automated (computer-assisted procedures) ordinarily use only two colors. Therefore, this chapter discusses only black symbols on a white background. This two-color system, however, can be extended to any two-color system, e.g., blueprinting, clear symbols on an opaque background, opaque symbols on a clear background or on a background of a contrasting color, etc.

Existing Standards

There are existing sets of standardized map symbols that are referenced in this chapter. The first, and most widely known, is the list of symbols for topographic maps published by the U.S. Geological Survey (USGS) (16). Though these symbols were designed primarily for use on maps of scale 1:10,000 and smaller, many are used on the maps of the scales being discussed in this chapter. The National Ocean Survey (NOS), a part of the National Oceanic and Atmospheric Administration (NOAA), with the coordination of the Defense Mapping Agency (DMA), publishes two pamphlets, which describe the symbols used on aeronautical charts (17) and nautical charts (4). These symbols, like the USGS symbols, are primarily used for small-scale maps, but may be used at the engineering scales this chapter analyses. The American National Standards Institute (ANSI) publishes the Y32.xx and the Z32.xx series publications (2). These are books of symbols used for special-purpose graphics, i.e., on plans for heating, ventilating, and air conditioning; plumbing and pipe fitting; valves and piping; etc. ANSI also publishes the book ANSI Y1.1, Abbreviations to be Used on Drawings and in Text (1). The spelling of geographical names, especially foreign names or names in other than the Roman alphabet, often poses a problem for map designers. The U.S. Board on Geographic Names, located at the USGS, Reston, VA, publishes standardized place names.

Lettering

Though names, abbreviations, and other alphanumeric identification are not considered as cartographic symbols, their contribution to the proper interpretation of the map is significant. Some conventions and guidelines for the placement and orientation of the lettering are presented in the following paragraph. The placement of the name is determined largely by the symbol location and the surrounding symbols. For point symbols, the preferred location of the lettering is to the right of and slightly above or below the center of the symbol. The positions of the lettering in the order of preference are shown in Fig. 2. Position 1 is the most preferred.

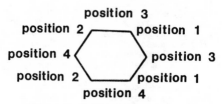

Figure 2 Positioning of Lettering

For linear features, the lettering should parallel the feature and use gentle curves to reflect the trend of the symbol (Fig. 3). Though the coastline is represented by a linear symbol, it generally is considered a boundary between two distinct areas. Names are usually perpendicular to the coastline and should not cross it (See Fig. 3). Names associated with areal features should be spaced across the area and appear well balanced within the area boundaries (Fig. 4). Lettering should not appear upside down, and should be readable without having to turn the map. Lettering along a near vertical line should be avoided. Where possible, lettering should parallel the upper and lower map borders. If lettering must cross a symbol, the symbol should be broken to allow the lettering to cross undisturbed. Fig. 4 provides examples of an area feature with lettering, a broken symbol with lettering, and a "clock" of lettering orientation (9,12).

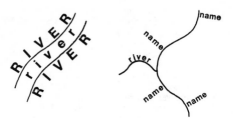

Figure 3 Orientation of Lettering

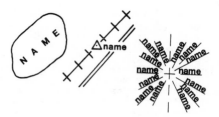

Figure 4 Name Depiction

Marginal Information

Marginal information is not often thought of as being map content. But, without the marginal notes the map would be almost useless, especially at the scales with which this chapter is concerned. As a minimum, the marginal information for engineering scale maps must include a title, the date of publication, a description of the location of the area that was mapped, the map scale, and a legend. Additional information that should be on a map includes: (1) The map projection; (2) the datum used; (3) a grid or graticule; (4) the direction of true north and magnetic north relative to the map's grid or graticule; (5) the contour interval; and (6) a description of the sources and accuracies of the data from which the map was compiled. A statement of the map's accuracy and reliability should also appear. The words "accuracy" and "reliability" as used in this chapter reflect the intent of Chapter 4, the chapter on map accuracies contained in this manual.

Map Content and Symbol Displacement

A map symbol is a graphic representation of a real world feature. The feature may be tangible like a building or abstract like a boundary line or grid. The degree of generalization on a map depends on the scale and purpose of the map. At 20 feet per inch (1:240), most features are drawn to scale. At 400 feet per inch (1:4,800), most features cannot be drawn to scale and are represented by a symbol. For example telephone poles, fire hydrants, railroad switches, guardrail posts, and manhole covers are all about the same size. If these are drawn to scale on a 1:4,800 scale map, they are 0.0025 in. (0.0635 mm) in diameter. Since human visual acuity at normal reading distances is about 0.004 in. (0.1 mm), the average person would be unable to discern the symbol. The symbol must be drawn larger than actual scale for the feature to be shown. If two adjacent features are to be represented, one symbol may have to be displaced from its original position on the map. If displacement must occur, a decision will have to be made as to which symbol to shift. Some guidelines to aid in making that decision are as follows:

1. Any point that is numerically defined by precise coordinates, i.e., a horizontal control point, must not be shifted on the map. (See Chapter 4, the chapter within this special publication on map accuracy, for the intent of the words "numerically defined" and "precise.")

2. Symbols should be located along the axis or centered over the location of the feature to the greatest extent possible.

3. If displacement must occur, symbols representing features pertinent to the purpose of the map should not be displaced. For example, in a map of the water distribution system of a city block, the symbol for a fire hydrant should be correctly positioned and the symbol for a lamp post should be displaced.

4. The displaced symbol should be placed as close as possible to its correct position, and shifted in the direction that best retains the

general relationship of the symbols involved, but least effects the adjacent symbols.

Relief Depiction

The purpose of a map may require a description of the third dimension, i.e., elevation. This chapter discusses only one method of describing the configuration of the surface -- contour lines. A contour line is a continuous line of equal elevation. Generally, every fifth contour line from the zero elevation line of the datum is called an index contour and is labelled with the elevation within the line. The index contour is usually drawn in a heavier line weight than the intermediate contour lines. A supplementary contour line may be used where the intermediate contour lines are too far apart for the surface configuration to be recognized. Supplementary contours are usually drawn as a dashed line of similar weight to the intermediate contours. Dashed lines also represent uncertain contours, e.g.,photogrammetrically derived contours under heavy timber. The depression contour is generally used to describe a small area that is of a lower elevation than the surrounding area, e.g., a sinkhole. It is normally a solid line of the same weight as the intermediate contour line, and it has small hatch marks perpendicular to the direction of the contour. These hatches are on the down slope side of the depression contour, and do not cross the contour line. Figure 5 shows examples of the previously mentioned lines.

A discussion of the contour interval is found in Chapter 3 of this special publication. Spot elevations are described in the section of this chapter titled, Symbols, Description and Use. Three conventions concerning the use of contour lines are: 1. The contour interval should remain constant on any one map unless the area mapped has both extremely steep and extremely flat terrain; 2. Contour lines should never cross except in rare cases such as Natural bridge; and 3. Contour lines should never split.

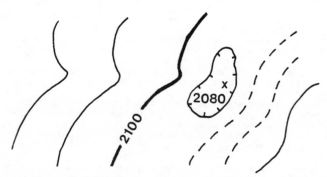

Figure 5 Relief Depiction (Contour Interval = 25 ft)

96

A problem that concerns convention number three is how to show a feature that is near vertical, such that the contour lines appear to run together or split. This is overcome by the use of a special symbol to depict the verticality. This symbol is discussed in the section, Symbols, Description and Use.

Development of the Symbols

A collection of 103 legends and lists of symbols was assembled. The symbol lists ranged from legends sent by small private firms through some based on international agreement. From this collection, tables of features were compiled with the symbols that represent each feature drawn in next to that feature. A single symbol for each feature was chosen from the tables. The criteria for symbol selection included: (1) Computer compatibility; (2) popularity among the users; and (3) utility at the scales of interest. If a suitable symbol could not be selected, a symbol was designed by the writer. The basic problem that was encountered was how to represent an existing feature, a proposed feature, and a destroyed, an abandoned, or an intermittent feature on the same manuscript. If the map is a planning map, the proposed structures should be emphasized. If the map is an operational or maintenance map, the existing features should be emphasized. Two solutions to this problem will be discussed in the next section.

Existing Versus Proposed Symbology

The title should specify that the map is displaying existing or proposed features as the primary purpose for the map's existence.

If the map is an operational map (see Fig. 6), then the existing features are shown as solid lines and closed or filled in symbols. The proposed features are shown as dashed lines and open or empty symbols. If the feature is an intermittent, abandoned, or destroyed feature, then the symbol has dots and dashes in the linear symbols, and the point symbols are broken or empty. The status of a feature, whether it is existing, proposed, or destroyed--, can be indicated by the symbol for the connecting linear feature as shown in figures 6 and 7.

Existing

Proposed

Destroyed

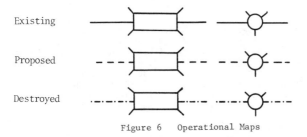

Figure 6   Operational Maps

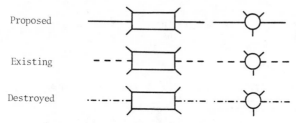

Proposed	
Existing	
Destroyed	

Figure 7 Planning Maps

For planning maps (see Figure 7), the abandoned, destroyed, or intermittent features are represented as they were on the operational maps. But, the proposed symbols are now represented as solid lines and closed or filled symbols, while existing features are represented by dashed lines and open or empty symbols.

A second solution to the problem of displaying existing versus proposed features is the use of overlays: Generally, transparent or translucent material is used, and each overlay or layer is prepared in a separate color. Any overlay or combination of overlays is possible, e.g., existing primary water distribution is shown in green, proposed primary water distribution in blue, existing local water distribution in red, and proposed local distribution in black, each on a separate overlay. Or, the entire water distribution system may be displayed on one layer in blue, using the system of symbols described in the preceding paragraphs. Then, the electric distribution system can be another overlay, the gas distribution system another, the phone system another, ad infinitum. Since much of the mapping at the engineering scales is done for planning and construction, urban development and control, and the maintenance and expansion of transportation systems for people and commodities, (see Chapter 3), the overlay system may greatly benefit the users of these kinds of maps. Using selected overlays, the maps can be tailored to specific requirements. A possible consideration to implementing the overlay system is the cost of production and storage.

Symbol Sources

If the map scale is larger than 20 feet per inch (1:240), it is recommended that the features be drawn to scale. If the map scale is smaller than 400 feet per inch (1:4,800), it is recommended that the map symbols published by the U.S. Geological Survey and the National Ocean Survey be used. For map scales between 20 feet per inch and 400 feet per inch (1:4,800), it is recommended that the following guidelines be used in selecting the map symbols. These guidelines may also be useful for selecting map symbols for smaller scale maps.

1. If the ANSI has a symbol for the feature that is compatible with the scale and the purpose of the map, that symbol should be used.

2. If ANSI does not have a symbol that is suitable, and a symbol listed in this chapter is acceptable, then that symbol should be used.

3. If the map contains specific features representing lights, buoys, beacons, radio/radar stations, or other special offshore/coastal features, then symbols from the previously described pamphlet, "Chart No. 1, Nautical Chart Symbols and Abbreviations" (4) should be used. If the map is to contain aeronautical information, e.g., radio aids to navigation, airway or controlled airspace information, symbols that warn aircraft of vertical obstructions, etc., then the previously described pamphlet, "Visual Aeronautical Chart Symbols" (17) should be referenced.

4. If symbols from none of the above sources or the USGS topographic map symbol list (16) meet the map designer's requirements, then a symbol developed by the map designer should be used. If the map designer develops a symbol, a copy of the symbol should be sent to the Committee on Cartographic Surveying of the Surveying and Mapping Division, ASCE, for publishing in the Journal of the Surveying and Mapping Division, and incorporation into the next edition of this special publication.

5. A legend or a statement of the sources of the symbols should be included on the map, e.g., "Symbols used on this map are from ANSI Y32.xx and the ASCE Special Publication on Map Uses, Scales, and Accuracies for Engineering and Associated Purposes."

Color

Though this chapter does not discuss color, if color is to be used on the map, it should conform to the USGS use of colors: blue for hydrography, brown for relief, green for vegetation, pink to designate built-up areas, red and yellow for filling in roads, and black for the cultural features, grids, and marginal information. If the map deals primarily with hydrography, the NOS or DMA use of colors may be more appropriate, e.g., shades of blue to represent depths.

Map Text

The ANSI publication, Y1.1, Abbreviations for Use on Drawings and in Text should be used to determine the correct abbreviation. The U.S. Board on Geographic Names publications should be consulted for the correct spelling of proper names. Guidelines for the orientation and positioning of lettering appears in the section of this chapter entitled, Lettering.

Map Titles

Titles should identify the purpose of the map and the area that is mapped. The title should identify the map as a planning or operational map, for example, "Proposed Storm Drainage System Map for the Northwest Quadrant of Hatboro."

Symbols, Description and Use

Tables 1-11, the tables of symbols, show for specific engineering applications the symbols as they should appear on an operational map. Labels and descriptions should be used with the symbols to avoid confusion and to identify correctly the feature that the symbol represents. In the tables, solid lines represent existing features, dashed lines represent proposed features, and dotted and dashed lines represent abandoned, destroyed, or intermittent features. Specific guidelines for using the symbols precede each table. Some general guidelines for using the symbols are:

1. Linear symbols should be centered on the manuscript location representing the centerline of the linear feature, e.g., road, railroad, river, powerline, gas line, etc., centerlines.

2. Features with intermittent support such as guardrails, aerial cableways, fences, hedgerows, etc., should be depicted with the post, pylon, trunk, etc., positioned on the manuscript to represent that feature's position and spacing. The line portions of the symbol should connect the point portions.

3. Point symbols that are symmetrical geometric figures, e.g., square, circle, rectangle, should be centered on the manuscript position that represents the location of the feature.

4. Symbols that represent terminating features, e.g., utility caps, plugs, anchors, should be perpendicular to the symbol representing the utility line, pipe, etc. The end of the linear symbol, not the point symbol, designates the end of the feature.

5. Odd shaped symbols, or symbols that are combinations of geometrical features, i.e., siren, toll booth, water fountain, ranger station, etc., have center points or centering circles as part of the symbol.

Transportation Symbols - Table 1.

1. Roads should be labelled with the number of lanes. Additional descriptive information, such as road width, bearing weight, construction, functional classification, etc., may also be used.

2. Multiple-track railroads should have the number of tracks shown or numerically described, depending on the scale of the map.

3. The symbols for point features, e.g., bridges, turntables, roundhouses, tunnels, etc., remain the same for existing, abandoned, or proposed structures. The status of the point feature (existing, proposed, etc.) is described by the status of the linear feature, i.e., road, railroad, etc., on which the point feature is located. If the status of the point feature is not the same as the status of the linear feature then the status of the point should be labelled. For example if a road exists, but the bridge is destroyed, or a railroad exists, but a

new bridge is being planned, the bridge should be labelled as destroyed
or planned. If both road and bridge are planned or exist, no label is
necessary; the status of the bridge is derived from the road.

TABLE 1 Transportation

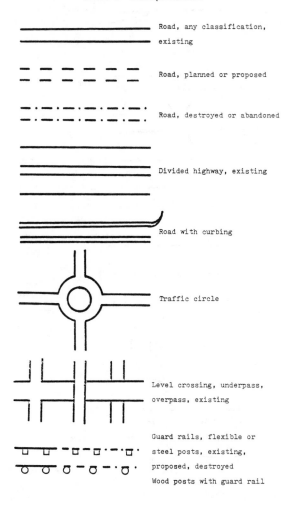

Road, any classification,
existing

Road, planned or proposed

Road, destroyed or abandoned

Divided highway, existing

Road with curbing

Traffic circle

Level crossing, underpass,
overpass, existing

Guard rails, flexible or
steel posts, existing,
proposed, destroyed
Wood posts with guard rail

TABLE 1 Transportation continued

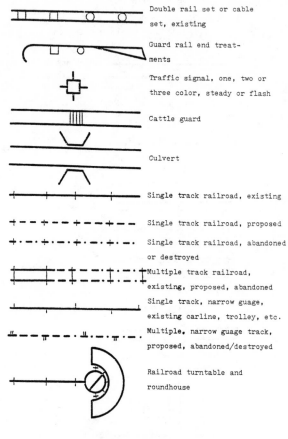

Double rail set or cable set, existing

Guard rail end treatments

Traffic signal, one, two or three color, steady or flash

Cattle guard

Culvert

Single track railroad, existing

Single track railroad, proposed

Single track railroad, abandoned or destroyed

Multiple track railroad, existing, proposed, abandoned

Single track, narrow guage, existing carline, trolley, etc.

Multiple, narrow guage track, proposed, abandoned/destroyed

Railroad turntable and roundhouse

Aerial cableway, tramway, cable supported transportation system, existing, proposed, and abandoned/destroyed

Ski lift, chair lift, existing, proposed, abandoned/destroyed

102

TABLE 1 Transportation continued

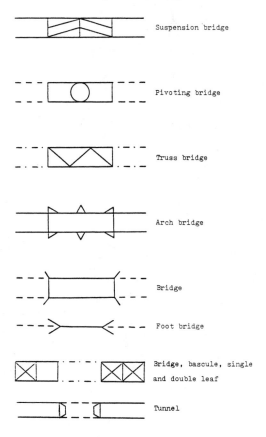

Suspension bridge

Pivoting bridge

Truss bridge

Arch bridge

Bridge

Foot bridge

Bridge, bascule, single
and double leaf

Tunnel

103

TABLE 1 Transportation continued

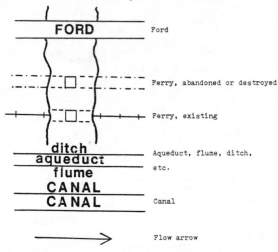

Ford

Ferry, abandoned or destroyed

Ferry, existing

Aqueduct, flume, ditch, etc.

Canal

Flow arrow

Utilities Symbols - Table 2.

   1. The cables, pipes, or other transportation methods will be
labelled as to what is being transported, and the size or capacity of
the transportation system, e.g., 6 inch Water Main, 200 psi Oxygen Line,
24,000 Volts, 3 in. Steam Line, Telephone, Cable TV, etc.

   2. As with the transportation symbology, there is only one symbol
for the point symbols representing utility features, not a separate
symbol for existing, for planned, etc. The status of the feature will
derive from the status of the transportation system symbol. For
example, if a fire hydrant is to be added into an existing water
distribution system, on the map, the fire hydrant symbol would be
labelled as "planned" or "proposed." Whereas, if the hydrant already
exists, the symbol for an existing water line would reflect the existing
status of the hydrant as well.

TABLE 2 Utilities

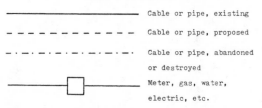

Cable or pipe, existing

Cable or pipe, proposed

Cable or pipe, abandoned
or destroyed

Meter, gas, water,
electric, etc.

104

TABLE 2   Utilities Continued

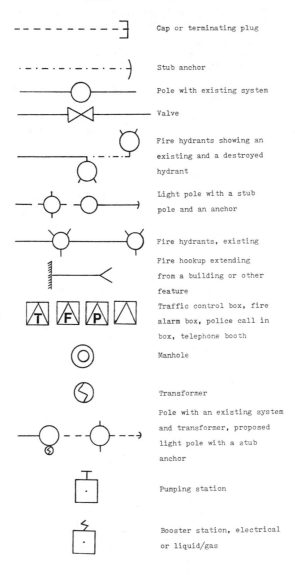

Cap or terminating plug

Stub anchor

Pole with existing system

Valve

Fire hydrants showing an existing and a destroyed hydrant

Light pole with a stub pole and an anchor

Fire hydrants, existing

Fire hookup extending from a building or other feature

Traffic control box, fire alarm box, police call in box, telephone booth

Manhole

Transformer

Pole with an existing system and transformer, proposed light pole with a stub anchor

Pumping station

Booster station, electrical or liquid/gas

Control Point Symbols - Table 3.

1. Control points should be labelled as to what order and class, and what agency established that point.

2. Other symbols are broken to allow the control point symbol to remain unbroken. Lettering should be placed so as to avoid overprinting a control point symbol.

3. Control points should be labelled as to type of survey; i.e., traverse, triangulation, vertical angle, etc.

4. If all the control points that are represented by the same symbol are of the same order and class, a statement may appear in the marginal notation describing the network, instead of labelling each point.

5. Spot elevations are labelled with the elevation.

6. The USGS is in the process of standardizing the symbology used on control diagrams.

TABLE 3   Control Points

△	Horizontal control point, national network
▽	Horizontal control point, state or local network
⊙	Located or landmark object
BM ✕	Monumented benchmark
BM △	Horizontal control point with benchmark
✕	Spot elevation
(railroad symbol)	Horizontal control point on a railroad, next to a road

Survey Line Symbols - Table 4.

Survey lines should be shown as a single solid line broken for labelling, i.e., centerline, property line, right-of-way line, etc. Survey lines should be labelled with sufficient information to identify the survey line or the two separated areas. If an abbreviation for the label is used, that abbreviation should appear in the legend.

TABLE 4   Survey Lines

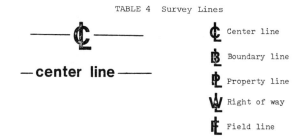

℄ Center line

ß Boundary line

℄ Property line

℄ Right of way

℄ Field line

Boundary Line Symbols - Table 5.

1. Boundary lines should be labelled with at least the names of the two adjoining areas, states, nations, etc. Only one kind of line will be shown, the line representing the highest jurisdiction.

2. The boundary line symbol designated as "other" should be labelled correctly as to its function, e.g., "National Wildlife Preserve."

TABLE 5   Boundary Lines

————— — — ————— — — —————   National

————— — ————— — ————— —   State/province

————— · ————— · ————— ·   Municipal

——— — ——— — ——— — ——— —   County, parish, etc.

——— ——— ——— ——— ———   Military

— — — — — — —   Other

107

Monument Symbols - Table 6.

Monuments should be labelled with sufficient information so that
they are correctly identified.

TABLE 6   Monuments

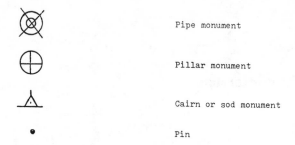

Pipe monument

Pillar monument

Cairn or sod monument

Pin

FENCE Symbols - Table 7.

1. Fences should be labelled with sufficient information to enable
the map user to identify them.

2. The gate symbol should be positioned such that the circle
designates the location of the hinges or the hinge support.

3. The rail, split rail, worm, etc., fence symbol should be
positioned such that the lower intersection designates the location of
the posts for fences with supporting posts.   (See Fig. 8)

TABLE 7   Fences

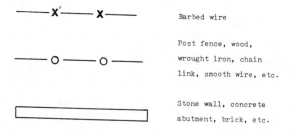

Barbed wire

Post fence, wood,
wrought iron, chain
link, smooth wire, etc.

Stone wall, concrete
abutment, brick, etc.

TABLE 7   Fences Continued

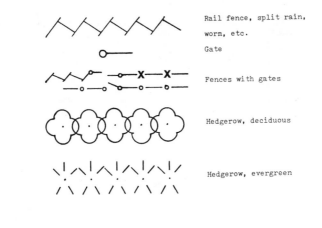

Rail fence, split rain, worm, etc.

Gate

Fences with gates

Hedgerow, deciduous

Hedgerow, evergreen

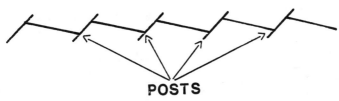

POSTS

Figure 8   Positioning of Split Rail Fence

Cultural Feature Symbols - Table 8A, B, and C.

Sufficient information to identify the feature should be provided, e.g., tanks should be labelled as to capacity, contents, etc. For example, a tank could be labelled: "60,000 gallons, #2 Crude Oil".

TABLE 8A Cultural Features, Miscellaneous

Catch basin, drop inlet, etc.

In sidewalk elevator

Sign and billboard

Siren, fire, air raid, hazardous weather, etc.

Light mounted on a building

Tower, microwave, electric transmission, radio or TV transmitter, etc.

Free standing pole, telephone, power, etc.

TABLE 8A Cultural Features, Miscellaneous continued

Flagpole

Gasoline pump island

Toll booth, tool shed, any
small booth or shed, exclu-
sive of telephone

Chimney

Tank, oil, water, above or
below surface, etc. should
be labelled

Cistern

Tree in a well, inside a
sidewalk, lawn, paved area,
etc.

TABLE 8B Cultural Features, Recreational

Mailbox

Water drinking fountain

Cave

Water mill or water wheel

Ranger station, park head-
quarters, park entrance,etc.

Wind mill, wind motor,
wind pump, etc.

111

TABLE 8B   Cultural Features, Recreational Continued

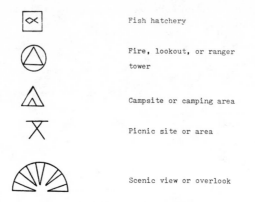

	Fish hatchery
	Fire, lookout, or ranger tower
	Campsite or camping area
	Picnic site or area
	Scenic view or overlook

TABLE 8C   Cultural Features, Mining

	Bore or test hole
	Distillate, oil, gas, etc., well, producing, dry, abandoned, etc. (label appropriately)
	Open pit mine or quarry, operational
	Open pit mine or quarry, abandoned, filled in, or no longer used.
	Mine, underground vertical shaft
	Mine, underground horizontal shaft

112

Hydrographic Feature Symbols - Table 9

    1. Rivers, streams, etc., will be bordered by an unbroken line, and be labelled with sufficient information for identification and status, e.g., Carol Creek (intermittent).

    2. Wells, springs, seeps, etc., will be labelled appropriately as to character: fresh, hot mineral, salt, artesian, etc.

TABLE 9 Hydrography

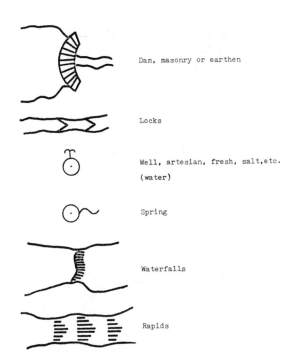

Dam, masonry or earthen

Locks

Well, artesian, fresh, salt,etc.
(water)

Spring

Waterfalls

Rapids

Surface Configuration (Topographic) Symbols - Table 10

    1. Contour lines, supplementary contours, intermediate contours, should be used to depict the configuration of the surface.

    2. The "natural vertical or near vertical slope" symbol indicates that the slope is too vertical to allow separate contour lines to be shown individually. This symbol preserves the convention that contour lines do not cross or split.

TABLE 10 Topography

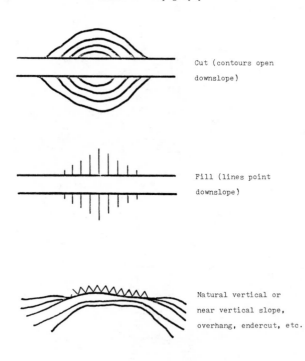

Cut (contours open
downslope)

Fill (lines point
downslope)

Natural vertical or
near vertical slope,
overhang, endercut, etc.

Vegetation Symbols - Table 11

1. These symbols were created to show individual, isolated, or landmark vegetation features. They can be used in conjunction with the boundary line symbol for "Other" to designate an area of special vegetation. Draw the boundary line symbol around the area to be identified, then randomly place the particular vegetation symbol within the area.

2. If the burned tree, cactus, or palm tree symbols are to be used to designate the location of point features, instead of a random pattern describing an area, the intersection of the lines depicting ground and trunk in the symbol represents the location of the feature.

3. For the stump symbol, the center point within the ellipse designates the location for an individual stump.

TABLE 11 Vegetation

Deciduous tree

Evergreen tree

Burned tree

Cactus

Stump, cut tree

Palm

Computerization

The set of symbols developed, in part, for this chapter was
designed with computer-assisted mapping methods in mind. The symbols
were developed for use on the Versatec, but may be used on most systems
that have move and draw functions. A comparison of computer software
for graphics is included in this chapter in Tables 12-19. The tables
tell what hardware the software will support, what languages the
software uses, what subroutines the software needs or uses, whether the
software package is two-dimensional or three-dimensional, what are the
host computers, what are the approximate costs, and lists some of the
advantages and disadvantages of each package.

TABLE 12 - DEFINITIONS

1. ADAGE Software

    The ADAGE graphics system is supported by the IMAGE FORTRAN
    extension and the GRAFX utility system. This system is fully
    interactive, supports a variety of ADAGE graphics terminals, and
    has been installed at many sites.

2. CalComp Software

    The widely used and much simulated CalComp software is a package
    exhibiting a large degree of host and output device independence.

3. DISSPLA

    The DISSPLA system is a complex set of graphing routines that
    provide extensive user-control for presentation of the images. It
    also enjoys a large degree of output device independence and has
    been implemented on a number of large-scale host machines.

4. GCS

    The Graphics Compatibility System (GCS) includes most of the
    features of the review criteria. The system includes graphs,
    segment control, 3D as well as 2D plotting, fairly complete input
    and control facilities, and has been implemented on many host
    machines at numerous sites with a variety of output devices.

5. GINO-F

    The GINO-F system is a set of routines that have, through
    evolution, become fairly host and device independent. The package
    is widely used in Europe, especially Great Britain, and the United
    States. It has found application in many computer aided design
    systems.

6. GPGS

    The General Purpose Graphic System (GPGS) is a structured and well-
    documented package that exists in both FORTRAN and host-dependent

versions. It has become the graphics standard in Norway and has, for several years, been the model for many proposed developments and comparisons. It also has been implemented and distributed widely.

7. IG

The Integrated Graphics (G) system has been developed recently at the University of Michigan. Though host-dependent, it enjoys a large degree of device independence. It contains many recent concepts in graphics systems, provides much capability, and uses other non-ANSII FORTRAN features to achieve convenient user interface.

8. Tektronix Software

The Tektronix Terminal Control System (TCS) is probably the most extensively used interactive graphics package—more than 2,000 installations. It is oriented as a graphing and plotting systems that is largely host independent and supports the various graphics I/O devices manufactured by Tektronix.

9. GSPC Core System

The GSPC Core System is prototype specification for a standard graphics subroutine package for use with line drawing plotters and displays. The design of the Core System was based on considerable experience with computer graphics software, and it embodies many concepts used in existing packages and some advanced concepts.

TABLE 13 Input Functions

	GRAPHICAL				INPUT DEVICES	TEXT	BUTTON
	PICKING		LOCATOR	VALUATOR			
	MENU	ITEM					
ADAGE	Returns menu index	Eight levels, address returned	2D, device units	1D, device units	Light pen, tablet, joystick/mouse,ball control dials	FORTRAN extensions	X
CALCOMP							
DISSPLA							
GCS	Returns menu index		2D, device or world units		Cursor, joystick, light pen, tablet, mouse, trackball	String of ASCII characters	
GINO-F		One level, segment no. returned	2D, device or world units	1D, device units	Installation dep.	String of ASCII characters	X
GPGS		n levels, segment no. returned	2D or 3D, device units	1D, device units	Clock, light pen, 1, 2, and 3D scalar devices	String of characters in AI format	X
GSPC		One level, segment no. returned	2D, device units	1D, user units		String of characters	X
IG		n levels, segment no. returned	2D, world units		Installation dep.		Inst. dep.
TEKTRONIX			2D, device or world units		Thumbwheels or joystick	String of ASCII characters	

118

TABLE 14 Transformations

	OBJECT TRANSFORMS				VIEWING TRANSFORMS				TRANSFORM MANIPULATION
	ROTATION	SCALING	TRANSLATION	OTHER	3D PROJECTION	WINDOWING	VIEWPORTING	SHIELDING	
ADAGE	2D 3D	2D 3D	2D 3D			2D 3D	Limited	X	Save,restore concatenate, nesting
CALCOMP		2D				2D (optional)			
DISSPLA	2D	2D	2D	User supplied	Parallel	2D 3D	X	X	
GCS	2D 3D	2D 3D	2D 3D		Perspective Parallel	2D 3D	X		Save and restore
GINO-F	2D 3D	2D 3D	2D 3D	2D & 3D shear	Perspective Parallel	2D 3D (limited)	Limited		Save,restore, concatenate, copy
GPGS	2D 3D	2D 3D	2D 3D	2D & 3D shear and user-supplied	Perspective Parallel	2D 3D	X		Save,restore concatenate
GSPC	Image Trans.	Image Trans.	Image Trans.	Image Trans.	Perspective Parallel	2D 3D	X		
IG	2D 3D	2D 3D	2D 3D		Perspective Parallel	2D	X		Concatenate, copy,nesting
TEKTRONIX	2D	2D				2D	X		

TABLE 15 Coordinate Systems and Picture Segments

	COORDINATE SYSTEMS			PICTURE SEGMENTS	
	WORLD	DEVICE	ORIENTATION	OPERATIONS	ATTRIBUTES
ADAGE	Fractional numbers from -1. to +1.	14-bit fractional numbers	Right	Open, invoke, transform, modify, close	Detectability, blink rate
CALCOMP	Inches or centimeters, extrema given by plotter size	Inches or centimeters, extrema given by plotter size	Right		
DISSPLA	Based on 8 1/2" x 11" page size, which may be reset	User does not deal with device units	Right		Specified for primitives
GCS	Range of floating point numbers	Raster units, inches, centimeters, percent, or character units	Right or left	Open, invoke, transform, delete, close	Specified for primitives
GINO-F	Range of floating point numbers	Millimeters	Right	Open, close, invoke, extend, delete, copy, rename, reposition	Detectability, visibility, blinking, intensity, color
GPGS	Range of floating point numbers or integers	Unit square	Right	Open, invoke, copy, close, delete	Visibility, intensity, blinking, color, depth cueing, detectability
GSPC	Not specified	Unit square	Right or left	Open, invoke, delete, rename, close	Detectability, visibility, highlighting, image trans.., retain
IG	Range of floating point numbers	Square from -1 to +1	Left	Open, extend, invoke, close delete, and transform	Intensity, hue, text size and font
TEKTRONIX	Range of floating point numbers	1024 x 1024 raster units	Right		

TABLE 16 Primitives

| | ATTRIBUTES | POSITION AND POINT GENERATION | PRIMITIVES | | |
			STRAIGHT LINES	CURVED LINES	SURFACES
ADAGE	Solid, dashed lines and text orientation	2D and 3D, absolute	2D and 3D, absolute	2D, circles and arcs	Contour and surface packages (Application software)
CALCOMP		2D, absolute	2D, absolute and polyline	2D, curves and arcs (Functional software)	
DISSPLA	Solid, four dashed, twelve arrowhead and user-defined lines	2D and 3D, absolute and multipoint plots	2D and 3D, absolute and polyline	2D and 3D, Several curve routines	Three methods for displaying surfaces
GCS	Color, intensity, blink, dashed lines, line terminators, text attributes	2D and 3D, absolute and relative	2D and 3D, absolute and relative	2D, circle, arc, and conic	
GINO-F	Solid, six dashed lines of specified color, width and type	2D and 3D, absolute and relative	2D and 3D, absolute and relative and polyline	2D and 3D, absolute and relative arcs and curves	
GPGS	Six line types	2D and 3D, absolute and relative	2D and 3D, absolute and relative and polyline	2D and 3D, absolute and relative arcs	
GSPC	Pick ID, color, intensity, line type, text attributes	2D and 3D, absolute and relative	2D and 3D, absolute and relative and polyline		
IG	Text size and font	2D and 3D absolute & relative positioning no point generation	2D and 3D, absolute and relative and polyline		
TEKTRONIX		2D, absolute and relative	2D, absolute and relative		

TABLE 16 Primitives Continued

| | | PRIMITIVES | | |
	MARKER	TEXT	NUMERIC OUTPUT	OTHER
ADAGE		2D transformable and 3D, several fonts, horizontal and vertical orientations	FORTRAN extensions	Raster output mode
CALCOMP	15 centered markers	2D transformable, one font	F, I formats	
DISSPLA	15 centered markers and user defined	2D and 3D transformable, thirteen alphabets and fonts	E, F, I formats	
GCS	Several markers and user defined	2D and 3D hardware and software transformable, two fonts	E, F, I formats	
GINO-F	8 markers	2D hardware and software transformable, two fonts	E, F, I formats	
GPGS	256 transformable markers	2D and 3D hardware and software transformable, two fonts	E, F, I formats	
GSPC	System- and user-defined markers	2D and 3D transformable, selectable fonts		
IG	Any character may be a marker	2D hardware or software transformable, four fonts	E, F, I formats	
TEKTRONIX	7 markers (AGII)	2D hardware, one font, four sizes	E, F, I formats (AGII)	

TABLE 17 Control

	INITIALIZATION	DEVICE CHARACTERISTICS	ERROR REPORTING	TERMINATION ROUTINE
ADAGE	None required	Set at system generation	Reports only display buffer overflow	X
CALCOMP	Single call to reserve buffer	Select a pen	Must be provided by user	X
DISSPLA	Call to initialize device Call to initialize common area	Select a "virtual" device	Complete error checking and reporting	X
GCS	Selects device and sets defaults	Set terminal mode, erase, flush buffer, manipulate status area, query status area	Complete error reporting to user	X
GINO-F	Select device	Sets device, plotter size, heading, speed, pen, erase, and query device	Error reporting to user	X
GPGS	Initialize device, initialize buffer	Select device, clear device, release device	Complete error checking and reporting, four severity levels	X
GSPC	Select core level	Select and initialize view surface, control updates, release devices	Error log or reporting to user	X
IG	Creates picture buffer, selects device	Select device, plotter size, heading, erase	Error reporting to user	
TEKTRONIX	Select terminal type and communication speed	Set terminal mode, erase, hardcopy, buffer type, flush buffer	None	X

123

TABLE Implementation Environment and Application Extensions

	IMPLEMENTATION ENVIRONMENT	APPLICATION EXTENSIONS		
		GRAPHING OPERATIONS	CURVE FITTING	POSITIONED TEXT
ADAGE	FORTRAN compiler with graphical extensions			Allows special menu generation text
CALCOMP	Coded in FORTRAN for larger machines, some assembler	Continuous, discrete, and 3D graphs supported by higher level	Spline and polynomial	
DISSPLA	Coded in FORTRAN	Continuous, discrete, and 3D graphs; shading provided	Spline, polynomial, linear interpolation	Allows many text display options
GCS	Coded in FORTRAN	Continuous and discrete graphs supported; 3D line plots	Spline, polynomial, linear interpolation	Allows text windows
GINO-F	Coded in FORTRAN	Continuous and discrete graphs supported by higher level package		
GPGS	ANSI FORTRAN, 360 Assembler, PDP11 MACRO			
GSPC	Seven installations investigating FORTRAN and APL implementations			
IG	360 Assembly language			
TEKTRONIX	Coded in FORTRAN	Continuous and discrete graphs supported by AGII		Allows text windows

TABLE 19 Miscellaneous Information

	NUMBER OF ROUTINES	GRAPHICS HARDWARE SUPPORTED	HOST COMPUTERS	NUMBER OF INSTALLATIONS	APPROXIMATE COST
ADAGE	IMAGE - 80 extensions GRAFX- 145 ops.	AGT/10, AGT/30, AGT/50, AGT/300 series	Adage DPR2 or DPR4 processors	87	Supplied with hardware
CALCOMP	Basic - 10 Functional - 43 Applications - 9 packages	All CalComp plotters and COM units	All major mainframes and several mini-computers	Several thousand	Basic software $500-$1,500
DISSPLA	206 routines	CalComp, CII, Xynetics, EAI, Tektronix, HP, Zeta, Broomall, & Gerber plotters; Tektronix, Computek, HI, VG, Adage, IBM CRTs; CalComp, III, and SC microfilm	Burroughs B6700, CDC 6000, 7000; IBM 360/370; DEC 10, 20 Honeywell 6000	100	$72,000
GCS	97 routines	CalComp, Gould, and TSP plotters; Computek and Tektronix storage tubes; IMLAC PDS1D and PDS4; line printer terminals	Burroughs 5700/6700/7700; CDC 6000; DEC 10; Honeywell 6000; IBM 360/370; UNIVAC 1100	118	Free
GINO-F	231 routines	Benson, Gradi, CalComp, ICL, Gerber, Kongsberg, Versatec, Zeta, CII, Laserscan plotters; Tektronix 4000 series; IMLAC, VG, DEC refresh; CADC raster device	Burroughs B6700, CDC 6600, DG Nova, DEC 10, 11, GEC, Honeywell 6000, IBM 370, ICL, Phillips 1400, Prime 300, UNIVAC 1100, Varian 620L, Xerox Sigma	120	$15,000
GPGS	116 routines	Tektronix 4000 series VG CalComp plotters	360/370 series PDP 11, PDP 15, UNIVAC 1108, Harris mini	40+	Assembler - $800 FORTRAN - $3000
GSPC	135 routines				
IG	35 routines	CalComp, HP, InkJet plotters; Tektronix, Computek, PDP storage tubes; Adage, DEC, IBM refresh devices; FR80, Pinto IV, and line printers; Diablo printers	IBM 360/370 AMDAHL 470	8	Free
TEKTRONIX	TCS - 84 AGII - 86	Tektronix 4000 series	All major mainframes and several mini-computers	Several thousand	TCS - $995 AGII - $850

125

Conclusion

The preceding recommendations together with the symbols in Tables 1-11 represent a start in an area that has been neglected for too long. Tables 1-11 are neither comprehensive nor complete. They are meant to be a foundation to be built upon. Standardization cannot take place overnight, or even in a year. It is a long-term process. As a base map is updated or redrafted, or as a new map is produced, it should include these recommendations as well as any that may be suggested in the future. In this way, eventually maps at these scales will become more easily interpreted by a wider selection of map users (9). "It would be presumptuous to imply that standardized graphic symbols will result in perfect intercommunication." (6). However, when an engineer from one office or state can pick up a map produced by another office or state, and use that map without becoming confused because the symbology is different, this chapter will have accomplished its purpose.

APPENDIX - REFERENCES

1. "Abbreviations to be Used on Drawings and in Text," ANSI Y1.1, American National Standards Institute, The American Society of Mechanical Engineers, New York, NY, 1972.

2. ANSI Series Y32.xx to Include Z32.2.3, Z32.2.4, Z32.2.6, American National Standards Institute, The American Society of Mechanical Engineers, New York, NY.

3. Board, C., "Cartographic Communication and Standardization, "International Yearbook of Cartography, Vol. 13, 1973, pp. 229-236.

4. "Chart No.1, Nautical Chart Symbols and Abbreviations," National Ocean Survey and Defense Mapping Agency, Department of Commerce, Washington, D.C., 1975.

5. Dornbach, J.E., "An Analysis of the Map as an Information System Display," thesis presented to Clark University, at Worcester, Mass., in 1967, in partial fulfillment of the requirements for the degree of Doctor of Philosophy.

6. Dreyfuss, H., Symbol Sourcebook, McGraw-Hill Book Co., Inc., New York, NY, 1972.

7. Feldscher, C.B., "A New Manual on Map Uses, Scales, and Accuracies," presented at the April 1-6, 1979, ASCE Convention and Exposition, held at Boston, Mass. (Preprint 3489).

8. "Final Report of the Graphic Standards Planning Committee, State of the Art Subcommittee," Computer Graphics, A Quarterly Report of SIGGRAPH-ACM, State of the Art Subcommittee, Graphic Standards Planning Committee, Vol. 12, No. 1-2, June 1978, pp. 14-33.

9. Jacober, R.P., Jr., "Standardization of Map Symbology for Large-Scale Maps," thesis presented to the Ohio State University, at

Columbus, Ohio, in 1979, in partial fulfillment of the requirements for the degree of Master of Science.

10. Keates, J.S., "Symbols and Meaning in Topographic Maps," International Yearbook of Cartography, Vol. 12, 1972, pp. 168-181.

11. Keates, J.S., Cartographic Design and Production, John Wiley and Sons, Inc., New York, NY, 1973.

12. Ormeling, F.J., Jr., "Procedures for Standardization of Cartographic Representation," presented at the July 26-August 2, 1978, International Cartographic Association IX International Conference on Cartography, held at College Park, MD.

13. "Report on Uniform Map Symbols, Parts I and II," Subcommittee on Uniform Map Symbols of the Committee on High Planning, The American Association of State Highway Officials, Washington, D.C., 1962.

14. Robinson, Sale, and Morrison, Elements of Cartography, John Wiley and Sons, Inc., New York, NY, 1978.

15. Steakley, J.E., "Large-Scale Mapping, Communication from Readers," American Cartographer, Vol. 4, No.1, Apr., 1977, p. 96.

16. "Topographic Maps," United States Geological Survey, Superintendent of Documents, U.S. Government Printing Office, Washington, D.C., 1976.

17. "Visual Aeronautical Chart Symbols," National Ocean Survey and Defense Mapping Agency, U.S. Department of Commerce, Washington, D.C., 1974.

18. Wilkins, E.B., "Maps for Planning," Public Administration Service, Chicago, IL, 1948.

CHAPTER 6. MAP AVAILABILITY

By Robert L. Brown[1] M. ASCE

INTRODUCTION

The planning and design of engineering works require some types of
mapping at the various stages of work. (See Chapters 2 and 3).
Initially, a small-scale map may be used for a general site selection,
as in the case of a regional airport. As the site selection process
narrows down the various alternatives, more detail becomes necessary,
and a larger-scale map would be helpful in locating utilities, and
interfacing with other modes of transportation.

Finally, as a final site is selected, a large-scale map showing
topographic details would be useful for designing runways, taxiways,
terminal buildings, hangars, parking areas, etc.

There are hundreds of different maps in circulation, each map
prepared with a specific usage intended. This chapter is intended to
alert the potential map user to the variety of maps suitable for
engineering projects, and planning purposes, which are available through
the public and private sectors of the business community.

United States Government Mapping

Listed in Appendix A are various map products produced by agencies
of the U.S. Government with information on the use and/or purpose of
each product. These products range from the popular topographic maps
published on a quadrangle basis by the U.S. Geological Survey (USGS) to
the lesser known statistical series of maps produced by the Bureau of
the Census. To assist the map user in selecting the specific map
products needed, various indexes are published by the agencies listed in
the appendix. These indexes include but are not limited to the
following: chart catalogs available from the National Ocean Survey
(NOS) and the Defense Mapping Agency (DMA), index maps and publication
lists from the U.S. Geological Survey (USGS), Corps of Engineers, U.S.
Forest Services (USFS), Bureau of Land Management (BLM), and state
highway departments and geological surveys.

The map user should also be aware of the existence of engineering
plans for various projects (dams, irrigation canals, roads, and other
developments) that may be available for purchase, usually at a nominal
fee. Such plans may be valuable for reference when considering (or
designing and constructing) adjacent developments.

[1]Mapping Supervisor, Colorado Department of Highways, Denver, Colorado.

State and City Mapping

There is also a considerable amount of mapping carried on at the
state government level. From a planning standpoint, one of the most
useful, albeit least known map products, is the General Highway Map
prepared by the state highway departments, although in some cases they
may be prepared by the office of the county engineer. The maps are
conveniently arranged to cover a given county in one or more sheets,
usually at the scale of 1 inch per mile (1:63,350 and 1:62,500). These
maps show the state and federal highway system, county road system,
drainage, railroads, pipelines, transmission lines, and some off-road
culture. In numerous cases, these maps have been used as a base map for
land use mapping, land ownership, origin-destination studies, and rural
addressing maps.

Other map products that are generally available from the majority
of the state highway departments are: (1) Traffic Volume/Traffic Flow
Maps; (2) Metropolitan Area Maps (also town, city, village, community
maps); and (3) State Maps.

All fifty states have a geological survey (or equivalent), many of
which may operate under a legislative mandate to conduct an ongoing
study or inventory of natural resources. A logical by-product of such
studies is a geologic map, report, or other form of earth science
information. Listed in Appendix II, for the fifty states plus the
District of Columbia, are the addresses of the State Department of
Highways (Dep't of Transportation or equivalent), and the State
Geological Survey or equivalent.

Three states have established the position of State Cartographer:
Colorado, South Carolina and Wisconsin. In a general way the office of
State Cartographer disseminates cartographic information within the
framework of the state government and provides technical assistance to
the local governments. It is to be noted that similar positions (or
duties) may or may not exist within the other state governments with
basically the same duties. The addresses of the state cartographer
offices are:

Colorado	State Cartographer Division of Local Government 1313 Sherman Street Denver, CO 80203
South Carolina	State Cartographer Geographic Statistics Office 915 Main Street Columbia, SC 29201
Wisconsin	State Cartographer 144 Science Hall University of Wisconsin Madison, WI 53706

The engineering offices of the larger cities generally have a variety of maps suitable for engineering and/or planning. The interested map user is referred to the appropriate office(s) of the municipal government.

Mapping By-Products

Modern mapping technology lends itself to the creation of various by-products of mapping that may be available from some agencies. These may include any or all of the following:

Aerial Photographs (Contact & Enlargements)

Photo Indexes and Line Indexes

Orthophotos & Orthophoto Quadrangles

Diapositives & Copy Negatives

Reproducible feature and/or color separates

Digital Tapes

The reproducible feature (or color) separates may be especially useful to persons needing only certain portions or features of a map, i.e., only the roads and drainage of a standard topographic map. The USGS intermediate-scale series are prepared in up to 25 feature separates. By combining only those features needed, the map user can design his/her own map.

National Cartographic Information Center

Having recognized the plethora of cartographic information produced by the mapping community, the National Cartographic Information Center (NCIC) was established by the USGS in 1974, replacing the Map Information Office and broading its function. The mission of NCIC is to disseminate information regarding cartographic data of the United States to the public and various Federal, State, and local government agencies. More than 30 Federal agencies gather and produce cartographic information. This includes maps and charts, aerial photography, geodetic control data and map data in digital form.

The USGS has embarked on a long-range objective in the establishment of a Digital Cartographic Data Base (3). Essentially all of the information shown on the existing 1:24,000 scale quadrangle maps will be included in this base. The data files may consist of digital elevation models, which are sampled arrays of elevations for a number of ground positions, or simply digital line graphs which are nothing more than line map information in digital form. In addition to the information from the 1:24,000 scale series of topographic maps for the Digital Cartographic Data Base, the USGS is responsible for storing, maintaining and disseminating to the users, the digital data from the 1:250,000 scale series of maps. This series of maps was digitized by the DMA originally for the production of raised relief maps (7).

Current applications of digital cartographic information include the generation of slope maps and terrain profiles. The interested map user is referred to the National Cartographic Information Center (NCIC) for more specific data on the Digital Cartographic Data Base. One should exercise some caution when considering the use of digital tapes, as some form of editing or "smoothing out" may have been performed by the originator of the tape. Should this conditon be suspected, the originator may be able to provide assistance such that tapes will not be ordered with required information having been edited from the tape.

Within the past decade, a program has been developed to obtain high-altitude quadrangle-centered aerial photography (2). Simply stated, this is photography flown such that every other photo is exposed over the center of a standard USGS 7 1/2-minute topographic quadrangle. The contact scale of the photography is 1:80,000 ±. At this scale, each quad center photograph covers the entire 7 1/2-minute quadrangle. By a ratioing process these aerial photographs can be enlarged to a scale which approximates that of the standard topographic quadrangle maps at 1:24,000 scale. In areas of minimal relief, photography can be enlarged without significant relief displacement. Where USGS coverage is available, (topographic maps) a standard practice is to rectify the photography to points on the 7 1/2-minute map. These photographs are the basis for producing orthphoto quads in areas of substantial relief. A significant development in this program is the proposal for a national high-altitude photographic base that would provide for cyclic coverage of the 48 conterminous states.

Hydrographic Surveys

Nautical charts provide the primary source of data for water and shoreline areas. A related product, the original plot of hydrographic surveys, provides more detailed information though generally over smaller areas. While not normally published, copies of these surveys can be obtained on request. Hydrographic surveys of the waters charted by the National Ocean Survey (formerly the U.S. Coast & Geodetic Survey and the U.S. Lake Survey) date back to the 1800's and cover the coastal waters of the United States, the Great Lakes, and adjacent waters. Basic data acquired in hydrographic surveys include: depths, bottom characteristics, currents, location of obstructions (rocks, wrecks & shoals) and aids to navigation (lights, buoys, daybeacons, landmarks). Topography of adjacent land areas is usually plotted on a separate drawing. Similar surveys are made by Districts of the Corps of Engineers, U.S. Army, and other organizations for charting, maintenance and construction of harbors and channels, flood control, power development, water supply, and water storage. In recent years the data have been acquired in digital form.

Highway Plans

Prior to the acquisition of right-of-way and eventual construction of a highway, detailed plans are prepared by or under the direction of the respective State Highway Department (or Dep't of Transportation). These plans show the new or proposed center line of the highway, existing culture and in some cases, the topography of the land by means of contours. The plans may be prepared at a variety of scales, the more common scales being 1:480, 1:600, 1:1,200, and 1:2,400. Although they are limited, some states use the scales of 1:240, 1:300, and 1:360.

Railroad Maps

In order to meet the requirements and interests of the individual railroads, and at the same time, satisfy the requirements of the Interstate Commerce Commission, the railroads maintain two series of maps

> Right-of-way and trackage maps

> Station maps

Right-of-way and track maps may be prepared at the scales of 1:1,200, 1:2,400 and 1:4,800, depending upon the amount of detail that needs to be shown. These maps show the following:

> Boundary lines of all rights-of-way

> Boundary lines of all detached lands

> Intersecting property lines of adjacent landowners

> Intersecting divisional land lines (Public Land Surveys)

> Division and subdivision of lands beyond the limits of the right-of-way

> Alignment and tracks

> Improvements

> Topographical features

Station maps are prepared at a scale of 1;1,200 or where greater detail is required, the scale is 1:600. (As used herein, the term station refers to a point designated for the origin or destination of freight). The station maps are prepared to show improvements in greater detail than possible on the right-of-way and track maps. All the details listed in the preceeding paragraph are shown on the station maps when they fall within the limits of the map (1).

Information regarding either series of the railroad maps may generally be obtained by contacting the individual railroads. Finally, one series of maps that would be useful, especially in the petroleum industry are the maps often referred to as "oil & gas maps." These maps are available from the private sector of business. Various editions of the maps provide base information, display maps and geologic structure maps. Coverage ranges from portions of a state (or region), to complete coverage of a state or region. The interested map user is referred to

132

the yellow pages of the telephone directory, under the general listing of "maps."

APPENDIX I. - VARIOUS MAP PRODUCTS

MAPS PUBLISHED BY THE BUREAU OF THE CENSUS

Name & Scale	Use/Purpose
State-County Subdivision Show the Maps	Developed from the 1970 census. subdivision of counties (for census purposes), and the location of all incorporated places and unincorporated places for which separate population figures are published.
U.S. County Outline Maps 1:5,000,000	Albers equal-area projection, sheet size 26" x 41
Urban Atlas	Individual maps, based on the 1970 census, for the 65 Standard Metropolitan Statistical Areas (SMSA), selected data characteristics. The maps are drawn by computer using micrographic mapping technology employing a chloropleth mapping technique to display data for each census tract within an entire SMSA.
Statistical Maps of the United States 1:5,000,000	Show the geographic distribution, by county, of various demographic and economic characteristics. Based on 1970 census data.
Metropolitan Map Series (MMS) 1:24,000	Cover the urbanized area portion of the majority of the SMSA's reported in the 1970 census. Show names of streets and other significant features within the area covered. Boundaries and names (or numbers) of places, Minor Civil Division (MCD), Census County Divisions (CCD), congressional districts, wards, census tracts, enumeration districts and blocks are shown. Grouping of blocks can also be ascertained from this series.
County Maps 1:125,000	This series shows those portions of metropolitan counties not covered by the MMS and the entirety of those counties outside of SMSA's. Boundaries of MCD's, CCD's, places, congressional districts, census

134

tracts, and Enumeration Districts (ED) are shown on this series except the ED's are not shown inside places for which place maps are available.

Place Maps
(various)

Place maps are based on maps supplied to the Bureau by local governments. This series shows streets and boundaries for places, MCD's, congressional districts, and enumeration districts. Also shown are census tracks, where applicable, and blocks if the place contracted with the Census Bureau for preparation of block statistics.

Tract Outline Maps
(various)

Show the boundaries, numbers or names of census tracts, counties, and all places with a population of 25,000 or more for the 241 metropolitan areas tracted in 1970. Only streets and map features which form tract boundaries are shown on this series of maps.

Urbanized Area Maps
1:250,000

Show the extent and components which make up the 279 urbanized areas defined for the 1970 Census with various shadings. A more detailed delineation of these boundaries can be found in the MMS.

Central Business District
Maps
(various)

Show the boundaries and identification of the census tracts that make up the CBD, and show street detail within the CBD as defined for the 1972 Census of Retail Trade.

Maps Published by the Census Bureau are generally available from:

Superintendent of Documents
Government Printing Office
Washington, D.C. 20402

MAPS AVAILABLE FROM THE BUREAU OF LAND MANAGEMENT (BLM)

Name & Scale   Use/Purpose

Intermediate Scale Maps   Published by the (BLM) for use in

1:100,000

management of the Public Lands and for other purposes. The maps in this series portray areas one degree of longitude by 30 minutes of latitude. Information shown includes townships, ranges, section lines, roads, drainage, towns and some cultural and physiographic features. The Surface Management Edition shows public lands administered by the BLM, other Federal Lands including those of the National Park Service, U.S. Fish & Wildlife Service, U.S Forest Service, State Lands, and Private Lands. Restrictions established by withdrawals are also shown. The Surface—Mineral Management Edition has, in addition to the preceding information, the Federally—owned Mineral Rights.

Cadastral Surveys
(various)

Cadastral Surveys of various units of the Public Lands. Includes township plats, mineral surveys, homestead entry surveys, and townsite surveys. Scale is dependent upon the area surveyed.

Special Mapping Projects
(various)

Each state office of the BLM is responsible for administering the public lands within its state. Therefore, each state office may have special mapping projects underway to meet its particular needs.

Information concerning maps published by the BLM may be obtained from the respective state offices, listed as follows:

Alaska
701 C. St., Box 13
Anchorage, AK 99513

Montana, North Dakota & South Dakota
222 N. 32nd St.
P.O. Box 30157
Billings, MT 59107

Arizona
2400 Valley Bank Center
Phoenix, AZ 99513

Nevada
Federal Building, Room 3008
Reno, NV 89509

California
Federal Bldg. Room E-2841
2800 Cottage Way

NM, OK, & TX
U.S. Post Office & Federal
    Building
P.O. Box 1449

136

Santa Fe, NM 87501

Colorado
1037 20th St.
Denver, CO 80202

Oregon & Washington
729 N.E. Oregon Street
P.O. Box 2965
Portland, OR 97208

States East of Mississippi
River, Iowa, Minnesota,
Missouri, Arkansas & Louisiana

Utah
University Club Bldg.
136 East South Temple
Salt Lake City, UT 84111

Eastern States Office
7981 Eastern Avenue
Silver Spring, MD 20910

Wyoming, Kansas & Nebraska
2515 Warren Avenue
P.O. Box 1828
Cheyenne, WY 82001

Idaho
Federal Bldg., Room 398
550 West Fort Street
P.O. Box 042
Boise, ID 83724

MAPS AVAILABLE FROM THE CORPS OF ENGINEERS

Name & Scale	Use/Purpose
Upper Mississippi River Navigation Charts 1:31,680	Navigations Rules & Regulations, Index Sheets, and Navigation Charts extending from the mouth of the Ohio River to Mile 868,10 miles above Minneapolis, Minn., and locating navigation aids and hazards, docks, recreation areas, historic sites, wildlife sanctuaries. Includes charts of the St. Croix River from the mouth to mile 26 above Stillwater, Minn. and 4 charts of the Minnesota River from the mouth to Mile 72, Shakopee, Minn. (Rock Island, St. Louis, St. Paul, Chicago)
Flood Control & Navigation Maps of the Miss. River; Cairo, Ill., to the Gulf of Mexico 1:62,500	Shows Mississippi River, with navigation sailing line, navigation aerial and submarine crossings, ferries roads, levees, and topography. (Vicksburg, St. Louis, Memphis, New Orleans)
Illinois Waterway Navigation Charts, Miss. Riv. at Grafton, Ill., to Lake Michigan at Chicago & Calumet Harbors	Covers both the Chicago Harbor route to Lake Michigan via the Chicago Sanitary and Ship Canal & the Chicago River, and the Sag

137

1:12,000	Route to Lake Michigan at Calumet Harbor via the Calumet-Sag Canal & Little Calumet & Calumet Rivers. Shows channel location, navigation aids, mileage, bridges, bridge clearances, normal sailing line, and the 9 foot contour within river banks. (Chicago, St. Louis)
Ohio River Navigation Charts; New Martinsville, West Virginia, to Pittsburgh, Pennsylvania 1:24,000	Shows sailing line, U.S. lights, day marks, arrival point markers for locks, normal pool elevation, tributary streams, location of bars, channel buoys, bridges, aerial crossings, seaplane bases, docks, terminals, landings, and navigation structures. (Pittsburgh)
Ohio River Navigation Charts; Foster, Kentucky to Martinsville, West Virginia 1:24,000	In addition to the information described in the preceding paragraph, these charts show the slack water pools adjacent to the Ohio River on the Little Kanawah River and The Big Sandy River. (Pittsburgh)
Ohio River Navigation Charts; Foster, Kentucky to Cairo, Illinois 1:24,000	Same information as described in preceding paragraph: New Martinsville West Virginia to Pittsburgh, Pennsylvania. (Louisville)
Green River Navigation Charts; 1:24,000	Shows sailing lines, critical bars, overhead structures, major tributaries, mileage above the mouth of the river and landmarks. Charts are made into a set from the mouth of the Green River at Mile 784.1 on the Ohio River, to the mouth of the Barren River. (Louisville)
Cumberland River Navigation Charts; Smithland, Kentucky to Mile 381.0 near Celina, Tennessee. 1:31,680	Shows sailing line, location of navigation aids, normal pool elevation, mouths of tributary streams, location of bars, bridges, aerial crossings, docks, terminals, landings, navigation structures and safety harbors. (Nashville)
Tennessee River Navigation Charts; Paducah, Kentucky to Knoxville, Tennessee.	Same information as described in preceding paragraph. (Nashville)

1:31,680

Missouri River Navigation Charts Kansas City, Missouri, to Sioux City, Iowa. 1:24,000	Charts show the normal sailing line river mileage, principal navigational aids, lights and daymarks, gages, bridges, overhead crossings with clearance data, dikes and revetments, adjacent towns and shore facilities (waterway freight terminals, way, docks, marinas, small boat docks, ramps, landings, and access roads), highways, and railroads. No channel depths, bars, or buoys are shown. (Kansas City, Omaha)
Missouri River Navigation Charts, Kansas City, Missouri, to Mouth 1;24,000	Same information as described in preceding paragraph. (Kansas City, Omaha)

Maps published by the Corps of Engineers are available from the respective District Offices, listed in parentheses, in the foregoing list. The addresses of the District Offices are as follows:

Chicago	219 South Dearborn St., Chicago, IL 60604
Huntington	P.O. Box 2127, Huntington, WV 25721
Kansas City	700 Federal Bldg., Kansas City, MO 64106
Little Rock	P.O. Box 867, Little Rock., AR 72203
Louisville	P.O. Box 59, Louisville, KY 40201
Memphis	668 Clifford Davis Federal Bldg., Memphis, TN 38103
Nashville	P.O. Box 1070, Nashville, TN 37202
New Orleans	P.O. Box 60267, New Orleans, LA 70160
Omaha	6014 U.S. Post Office & Court House, 215 North 17th St., Omaha, NE 68102
Pittsburgh	Federal Bldg., 1000 Liberty Ave., Pittsburgh, PA 15222
Rock Island	Clock Tower Bldg., Rock Island, IL 61201
St. Louis	210 North 12th St., St. Louis, MO 63101
St. Paul	1135 U.S. Post Office & Custom House, St. Paul, MN 55101

Tulsa   P.O. Box 61, Tulsa, OK 74102

Vicksburg   P.O. Box 60, Vicksburg, MS 39180

MAPS AVAILABLE FROM THE INTERNATIONAL BOUNDARY COMMISSION

The International Boundary Commission, U.S. & Canada, United States Section, has available a series of reports and maps covering the survey and demarcation of the International Boundary between the United States and Canada.

The reports are listed thus:

Joint Report Upon the Survey and Demarcation of the International Boundary Between the United States & Canada:

1. Along the 141st Meridian from the Arctic Ocean to Mt. St. Elias (1918)

2. From the 49th Parallel to the Pacific Ocean (1921)

3. From the source of the St. Croix River to the St. Lawrence River (1924)

4. From the Northwesternmost Point of Lake of the Woods to Lake Superior (1931)

5. From the source of the St. Croix River to the Atlantic Ocean (1934)

6. From the Gulf of Georgia to the Northwesternmost Point of Lake of the Woods (1937)

7. From Tongass Passage to Mount St. Elias (1952)

Available from:

International Boundary Commission
United States & Canada
425 I Street, N.W., Room 150
Washington, D.C. 20536

The International Boundary Commission, United States and Mexico, United States Section, has compiled a photomosaic map showing the International Boundary with Mexico in the Colorado and the Rio Grande Rivers. The photography for the Rio Grande was taken in 1972 and 1974; for the Colorado River, 1973 and 1975.

The scale of the maps is 1:20,000 with Geographic, Universal Transverse Mercator (Zone 14) and Texas (South Zone) State Plane Coordinates. The map of the Rio Grande consists of 94 sheets plus title sheet and two indexes. The map of the Colorado River consists of three sheets plus title sheet and index sheet.

Available from:

Branch of Distribution
U.S. Geological Survey
Box 25286
Denver Federal Center
Denver, CO   80225

MAPS AVAILABLE FROM THE NATIONAL OCEAN SURVEY

Aeronautical Charts (4)

Name & Scale	Use/Purpose

Sectional Charts
Sectional: 1:500,000
Local: 1:250,000

Designed for visual navigation of slow/medium speed aircraft. Topographic information features the portrayal of relief and judicious selection of visual check points such as populated places, drainage, roads, railroads, and other distinctive landmarks suitable for Visual Flight Rules (VFR) flight. Aeronautical information includes visual and radio aids to navigation, aerodromes, controlled airspaces, restricted areas, obstructions and related data.

VFR Terminal Area Chart
1:250,000

Same as Sectional Charts, in addition to depicting the airspace designated as "terminal control area," which provides for the control or segregation of all aircraft within the terminal control area. Aeronautical information is identical to that shown on Sectional Charts.

World Aeronautical Chart
1:1,000,000

Standard series of aeronautical charts covering land areas of the world, at a size and scale convenient for navigation by moderate speed aircraft. Topographic information includes cities and towns, principal roads, railroads, distinctive landmarks, drainage, and relief. Aeronautical information includes visual and radio aids to navigation, aerodromes, airways, restricted areas, obstructions and other pertinent data.

Enroute Low Altitude Charts

Enroute Low Altitude Charts provide aeronautical information for enroute navigation in the low altitude stratum. Information includes the portrayal of Low & Medium Frequency (L/MF) and Very

High Frequency (VHF) airways; limits of controlled airspace; position, identification and frequencies of radio aids; selected aerodromes; minimum enroute and obstruction clearance altitudes; airway distances; reporting points; special use airspace areas; and related information.

Enroute High Altitude Charts

Enroute High Altitude Charts provide aeronautical information for enroute instrument navigation in the high altitude stratum. Information includes the portrayal of jet routes, position, identification and frequencies of radio aids; selected aerodromes; distances; time zones; special use airspace areas; and related information.

Area Navigation High Altitude Charts (RNAV)

RNAV High Altitude Charts provide aeronautical information for air routes established for aircraft equipped with area navigation systems. Information includes portrayal of RNAV routes, way-points, track angles, changeover points, distances, selected navigational aids and aerodromes, special use airspace areas, oceanic routes, and other transitional information.

Aircraft Position Charts
1:5,000,000 to
1:6,750,000

Suitable for plotting lines of position from celestial observation and electronic aids. Designed for long-range flights usually over extensive water or desert areas. Topographic information includes spot elevations, important drainage features, large cities, and international boundaries. Aeronautical information for overwater flights includes oceanic control areas, flight information regions, weather zones, normal positions of air-sea rescue vessels, and loran and consol data, if required.

VFR/IFR Planning Chart
1:2,333,232

Designed to fulfill the requirements of preflight planning

143

for Visual Flight Rules (VFR) and
Instrument Flight Rules (IFR)
operations. Printed front/back
such that when assembled, a
composite VFR Planning Chart is on
one side, and IFR Planning Chart on
the other. IFR Chart includes low-
altitude airways and mileage,
navigational facilities, special-
use airspace areas, time zones,
airports, isogonic lines, and
related data. VFR Chart includes
selected populated places, large
bodies of water, major drainage,
shaded relief, navigational
facilities, airports, special-use
airspace areas, and related data.

Flight Case Planning Chart
1:4,374,803

This chart is designed for preflight planning and enroute flight
planning for Visual Flight Rules
(VFR). Contains basically the same
information as the VFR/IFR Planning
Chart (see preceding listing) with
the addition of selected Flight
Service Stations (FSS) and Weather
Service Offices located at airport
sites, parachute jumping areas, a
tabulation of special use airspace
areas, a mileage table listing
distances between 174 major
airports and a city/aerodrome
location index.

Gulf of Mexico & Caribbean
Planning Chart
1:6,270,551

This chart is designed for preflight planning for Visual Flight
Rules (VFR) operations. This chart
is the reverse of the Puerto Rico -
Virgin Islands Local. Information
includes mileages between Airports
of Entry, a selection of special
use airspace areas, and a Directory
of Aerodromes with their available
facilities and servicing.

Jet Navigation Charts
1:2,000,000

To provide charts suitable for long-
range high speed navigation.
Topographic features include large
cities, roads, railroads, drainage,
and relief. The latter is
indicated by contours, spot
elevations and gradient tints.
Aeronautical information includes
restricted areas, Low/Medium

Frequency (L/MF) and Visual Omni
Ranges (VOR), radiobeacons, and a
selection of standards broadcasting
stations and aerodromes.

Global Navigation Charts
1:5,000,000

To provide charts suitable for
aeronautical planning, operations
over long distances and enroute
navigation in long-range, high-
altitude, high-speed aircraft.
Topographic information includes
cities, towns, drainage, primary
roads and railroads, prominent
culture, and shaded relief
augmented with tints and spot
elevations. Aeronautical data
includes radio aids to navigation,
aerodrome and restricted areas.

Global LORAN Navigation Charts

Same scale and coverage as the
Global Navigation Charts.
Topographic information includes
major cities only, coastlines,
major lakes and rivers, and land
tint. No relief or vegetation.
Aeronautical data includes radio
aids to navigation and Long-Range
Navigation (LORAN) lines of
position.

Airport Obstruction Charts
1:12,000

These charts show runways and
flight paths for landing and take-
off, together with the positions
and elevations of the objects which
are potential hazards to these
operations. Charts are used in
determining the maximum safe take-
off and landing gross weight of
civil aircraft; in determining
airport instrument approach and
landing procedures, and relative to
clearing of obstructions and
improvement of airport facilities.

Available from:

Distribution Division (C44)
National Ocean Survey
Riverdale, MD   20840

Geodetic Diagrams

Name & Scale	Use/Purpose
State Horiz. Control Diagrams 1:500,000 to 1:666,000	Depict the approximate location of horizontal control marks by State, combined States, or portions of States. Primarily used for planning purposes.
State Vertical Control Diagrams 1:500,000 to 1:666,000	Depict the approximate location of vertical control level lines by State, combined States, or portions of States. Primarily used for planning purposes.
Geodetic Control Diagrams 1:250,000	Depict by 1° x 2° quadrangles, vertical and horizontal control data established by NOAA (NOS), USGS, and other contributing agencies. Used for the selection of geodetic control data and for the planning of national, State, and local surveying and mapping projects.
Nautical Diagrams (Various)	Charts for the coastal areas of Alaska, Hawaii and the contiguous U.S. coastal areas; these diagrams depict the location of horizontal control marks and in some cases, vertical control level lines. Primarily used for coastal and offshore surveying and mapping operations.
Aeronautical Diagrams 1:500,000 to 1:5,000,000	Networks depicted on charts of Alaska show the location of horizontal control marks and the vertical control level lines. Used for project planning and selection of geodetic control data.
Urban Diagrams (various)	Cities and inserts for areas containing very dense control; show the location of horizontal control marks and/or vertical control level lines. Used primarily for local or State surveying and mapping operations.
Vertical Control Quad Diagrams 1:250,000	Depict the direction and location of level lines occurring in a

particular geographically defined area, and serve as pictorial indices to assist the map users in selection of vertical control data for specific areas.

Special Purpose Control Diagrams: St. Lawrence River Diagram	Depicts the location of horizontal control marks along the river(published by NOAA and International Boundary Commission, U.S. and Canada); used for planning purposes.
District of Columbia Diagram 1:60,000 Nominal Scale	Depicts location of horizontal control marks within the District of Columbia (published by NOAA and the Federal Highway Administration (FHWA); used for planning purposes.
Horizontal Control Progress Sketches (various)	Available upon special request; these diagrams are used to: (1) supplement previously published horizontal control diagrams prior to diagram revision and republication; (2) complement field survey data for project planning and network adjustments; and (3) research epochal developments of national horizontal network in specific areas.

Available from:

National Geodetic Information Center, N/CG17
National Geodetic Survey
NOAA/NOS
Rockville, MD 20852

Nautical Charts

Name & Scale	Use/Purpose
Sailing Charts 1:600,000 and smaller	Plotting charts for offshore sailing between distant coastal ports and for approaching the coast from the open ocean.
General Charts 1:150,000 to 1:600,000	Offshore navigation, when position can be fixed by landmarks, lights, buoys, and characteristic soundings.

Coast Charts 1:50,000 to 1:150,000	Coastwise navigation inside the offshore reefs and shoals, entering bays and harbors of considerable size and navigating certain inland waterways.
Harbor Charts 1:50,000 & larger	Navigation and anchorage in harbors and small waterways.
Intracoastal Waterway Charts 1:40,000	Navigation within the Intracoastal Waterway.
Small Craft Charts 1,80,000 & larger	Published with small craft information.
Marine Boundary Maps & Charts (various)	Portray the Territorial Sea (3-mile limit) and contiguous zone (12-mile limit) and/or Fishery conservation Zone (200 mile limit) in U.S. waters.
Offshore Mineral Leasing Area Maps (various)	Shows offshore Mineral Leasing Areas and blocks overprinted from data furnished by Bureau of Land Management (BLM). Not intended for navigation or for official lease block information, but can be used as a reference when navigating in the offshore lease area.
Storm Evacuation Maps 1:62,500	Designed to show areas that would be inundated by various levels of hurricane generated storm surges, and the best route for evacuation inland. Used by federal, state and local officials for hurricane preparedness planning and are available to the general public.
Geophysical Maps (various)	Each consists of three sheets, (a base bathymetric map, a magnetic map and a gravity map), and where practicable, a sediment overprint (NOS 1308N-17SO). The bathymetric maps serve as a base for making geological-geophysical studies of the ocean bottom's crustal structure and its composition. The 1:250,000 series contains the geophysical data for the continental shelf and slope. The Sea-map Series, at a scale of 1:1,000,000 covers geophysical data

gathered in the deep sea area, sometimes including the adjacent continental shelf and slope.

Florida Coastal Zone Maps
1:10,000 Orthophoto

Provides data for the selection of baseline points to establish coastal boundaries, including seaward boundaries and between sovereign land and the uplands, subject to private ownership for the State of Florida. These maps show tidal datum lines delineated from the tide-coordinated infrared photography, non-floating aids to navigation, landmarks and other significant features. Produced in cooperation with State of Florida, Department of Natural Resources.

Bathymetric Charts
(various)

Topographic maps of the sea floor. Through the use of detailed depth contours and full use of bathymetric data, the size, shape and distribution of underwater features are vividly portrayed. The Bathymetric Map serves as the basic tool for performing scientific, engineering, marine geophysical and marine environmental studies, that are required in the development of energy and marine resources. Some maps available with BLM Outer Continental Shelf (OCS) lease block area.

Topo/Bathy Maps
1:250,000

Multipurpose maps show the National Ocean Survey bathymetry (ocean bottom topography) and the U.S. Geological Survey (USGS) land topographic information. Useful for land-use planners, conservationists, oceanographers, marine geologists, and others having an interest in the coastal zone and the Outer Continental Shelf's (OCS) physical environment. Overprinted with BLM OCS lease blocks. Cooperatively produced by NOS & USGS in support of the Coastal Zone Management and Energy Impact Programs and the offshore oil and gas program.

Tidal Current Charts

Each of these publications has a set of 11 pages that depict by means of arrows and figures, the direction and velocity of the tidal current for each hour of the tidal cycle. The charts which may be used for any year, present comprehensive views of the tidal current movements in the respective waterways as a whole and also supply a means for readily determining, for any time, the direction and velocity of the current at various localities throughout the water areas covered. The Narraganset Bay and New York Harbor tidal current charts are to be used with the annual tide tables. The other charts require the current tables.

Available from:

Distribution Division (C44)
National Ocean Survey
Riverdale, MD 20840

Original Hydrographic and Topographic Surveys

Planetable and, in later years, photogrammetric surveys of the coastal area, and hydrographic surveys of the adjacent waters have been in progress by the National Ocean Survey (formerly U.S. Coast & Geodetic Survey and U.S. Lake Survey) for the production and maintenance of nautical charts since 1835. As a result of this activity, over 23,000 individual surveys are on file in the National Ocean Survey archives.

These represent a unique and comprehensive record of our coastline and the adjacent water, including the Great Lakes, showing conditions existing at a particular date over more than a century, and providing a quite detailed record of the changes that have occurred from both natural and man-made causes. These records are used extensively by the public and other agencies of the government for research, engineering and development purposes.

Topographic and hydrographic surveys are made and recorded separately. For studies of water depths, soundings, etc., hydrographic surveys should be requested; for studies of the shoreline and adjacent land areas, topographic surveys should be requested. Topographic surveys vary in coverage and content. Many show only the shoreline and planimetric features immediately adjacent thereto. Others are complete topographic maps covering from the shoreline inland for as much as five or more miles or up to the 71/2-minute quadrangle limit.

Available from:

National Ocean Survey
Data Control Branch (C353)
Rockville, MD 20852

MAPS AVAILABLE FROM THE SOIL CONSERVATION SERVICE

Name and Scale	Use/Purpose
Soil Surveys 1:15,840 to 1:31,680	To show distribution of soil types. Usually accompanied by text guide for interpretation.
Prime Farmlands Maps 1:50,000 or 1:100,000	Show unique, prime and important farmland.
Watershed 1:800 to 1:126,720	To delineate watershed boundary, structure with drainage area, benefited area.
River Basins 1:10,000 to 1:500,000	To show limits of study areas, and geographic distribution of basin features.
Resource Conservation & Development (RC&D) 1:24,000 to 1:100,000	Various thematic resource data maps mainly for use by local sponsor in developing an economic plan for an area.
SCS Activities Map 1:5,000,000	Administrative areas, Districts, RC&D areas, Watersheds & Great Plains Program.
Soil Survey Status Map 1:5,000,000	Shows counties that have published surveys, surveys completed but not published, and surveys in progress.
Special Maps on Soils and other Resources (various)	Resources are shown for planning purposes by units of local government.
Flood Hazard Maps 1:4,800 to 1:6,000	Shows delineation of flood frequencies.

Available from:

Soil Conservation Service
U.S. Department of Agriculture
Washington, D.C. 20015

Note: many of the maps are
developed primarily to satisfy the

on-going requirements of the SCS
programs. The user is urged to
contact SCS at the above address
prior to ordering any maps. The
Soil Survey Series are generally
available from:

Government Printing Office
Washington, D.C.   20402

MAPS AVAILABLE FROM THE TENNESSEE VALLEY AUTHORITY (TVA)

Name & Scale	Use/Purpose
Topographic Quadrangle 1:24,000	7 1/2-minute series, by TVA-USGS. (Map content is identical to that listed under USGS Nat'l Topographic Map Series). Coverage is limited to the Tennessee Valley Watershed.
Charts of Tennessee River River Waterway 1:62,500 & 1:31,680	Navigation charts showing the main river channel sailing line, safety harbors, secondary channels, navigation aids (including daymarkers), general shape and elevation of lake bottom. Underwater areas containing stumps, and other map features such as drainage, woods, roads, buildings, and boundaries.
Navigation-Recreation Maps of Tributary Reservoirs 1:12,000 & 1:31,680	Navigation charts showing navigation markers, recreation information (launching ramps, docks, and campgrounds), contours, and topographic information (transportation systems, cultural facilities, drainage, and contours) for surrounding lands.
Reservation Property Maps (various)	Reservation maps showing TVA boundary line data and adjacent property corners.
Reservoir Area Maps (various)	Reservoir area maps revised since 1970 showing reservoir and surrounding region, drainage, highways, railroads, and shorelines of lakes at normal pool elevation.

Available from:

Tennessee Valley Authority
Map Information and Records Unit
Chattanooga, TN 37401

or

Tennessee Valley Authority
Maps and Engineering Records
Section
Knoxville, TN 37902

MAPS AVAILABLE FROM THE U.S. FOREST SERVICE

Name & Scale	Use/Purpose
Primary Base Series 1:24,000	Information content is identical to the National Topographic Map series with the exception of the following added information: Forest roads and trails, with identification numbers (names when appropriate); private land holdings within the national forest; Ranger District Boundaries; and additional information essential for forest management.
Secondary Base Series 1:126,720	Commonly referred to as the Forest Visitor Maps. Features shown include State and U.S. Highway systems, forest roads and trails, other transportation facilities, schools, buildings, horizontal and vertical control stations, Forest Service Ranger & Guard Stations, hydrographic information and recreation information.
Travel Map 1:126,720 & 1:168,960	Overprint of Secondary Base Series, described in preceding paragraph, that shows areas, roads and trails that are either closed to Off-Road Vehicle (ORV) use, or have restrictions on the use of ORVs. Also covers restrictions governing motorized trail vehicles.

Information concerning maps published by the Forest Service may be obtained from the Regional Offices, listed as follows:

Forest Service, USDA
Northern Region
Federal Building
Missoula, MT 59807

Rocky Mountain Region
11177 West 8th Avenue
P.O. Box 25127
Lakewood, CO 80225

Southwestern Region
Federal Building
517 Gold Ave., S.W.
Albuquerque, NM 87102

Pacific Northwest region
319 SW Pine Street
P.O. Box 3623
Portland, OR 97208

Southern Region
1720 Peachtree Road N.W.
Atlanta, GA 30309

Eastern Region
633 West Wisconsin Avenue
Milwaukee, WI 53202

Intermountain Region
324 25th Street
Ogden, UT 84401

Alaska Region
Federal Office Building
P.O. Box 1628
Juneau, AK 99802

California Region
630 Sansome Street
San Francisco, CA. 94111

MAPS AVAILABLE FROM THE U.S. GEOLOGICAL SURVEY (5)

Name & Scale	Use/Purpose
National Topographic Map Series 1:24,000	Quadrangle maps of the U.S. covering 7 1/2-minutes of latitude and longitude, showing the network of hydrographic features, networks of roads, highways, railroads, and other cultural features and the shape and elevation of the land and landforms. Used as base maps for Geologic Maps and Hydrologic Maps.
Geological Maps (various)	In addition to the information listed in the preceding paragraph, geological maps show the characteristics and distribution of rocks and surficial materials by age and their physical and structural relation with one another.
Hydrologic Maps 1:24,000	In addition to the information listed in the preceding paragraph under Nat'l. Topographic Map Series, hydrologic maps portray the occurrence of water on and beneath the land surface. Flood prone area maps approximate the boundaries of the "100 year flood."
Intermediate Scale Maps 1:50,000 & 1:100,000	Relatively new (1975) series of topographic maps. Intended to provide for additional levels of detailed information between the 1:24,000 and 1:250,000 scales. These series are produced in feature-separation format. Using as many as 25 separate feature plates, it is possible to combine any number of plates and end up with a map containing only those features wanted or needed. In some areas, maps compiled on a county

	format at the 1:50,000 scale are available.
1:50,000, 15 min. quadrangles	Jointly produced by the U.S.G.S. and the Defense Mapping Agency (DMA). Prepared from existing 7 1/2-minute quadrangles, with metric contours and feature separation techniques. Coverage is for selected areas of the U.S.
1:250,000 Maps of the United States	This map series represents the only series with complete coverage of the entire United States. Format is one degree of latitude by two degrees of longitude. Contour intervals range from 50 feet for relatively flat areas, to 200 feet in mountainous areas. Originally prepared by the Army Map Service, they are now maintained by the U.S.G.S.
State Base Maps 1:500,000	Generally published in three editions (1) Base Map; (2) Highway and Contour Map; (3) Shaded Relief Map. Features shown on Base Maps include county boundaries, cities, towns, villages, settlements, railroads and water features. Highway and Contour Maps are overprints of the Base Map with highways, contours, national parks, forests, monuments, and wildlife refuges added. Shaded Relief Maps show county boundaries, larger cities, and water features. Also available in half-scale (1:1,000,000) in black and white, Base Map only.
River Survey Maps (various)	These maps show the course and fall of the stream, configuration of the valley floor and adjacent slopes, and location of towns, scattered houses, irrigation ditches, roads and other cultural features. Potential reservoir sites are mapped to the probable flow line of the reservoir. Some maps include proposed dam sites on a larger scale and include a profile of the stream. These maps were prepared in connection with the

156

classification of the public lands,
and consequently cover areas in the
Western States.

Maps published by the U.S. Geological survey are available from
some commercial firms in the various states, and from two main
distribution centers:

1.   For areas east of the Mississippi River, including
     Minnesota, Puerto Rico and the Virgin Islands:

     Branch of Distribution
     U.S. Geological Survey
     1200 South Eads Street
     Arlington, VA 22202

2.   For areas west of the Mississippi River, including Alaska,
     Hawaii, Louisiana, American Samoa, and Guam:

     Branch of Distribution
     U.S. Geological Survey
     Box 25286
     Denver Federal Center
     Denver, CO 80225

APPENDIX II - INFORMATION SOURCES FOR STATE MAPPING

Alabama   State of Alabama Highway Dept.   Geological Survey of
                 State Highway Building   Alabama
                 Montgomery, AL 31630   P.O. Drawer O
                                                     University, AL 35486

Alaska   Alaska Dept. of Transportation   Div. of Geological &
                 P.O. Box 1467   Geophysical Surveys
                 Juneau, AK 99802   3001 Porcupine Drive
                                                     Anchorage, AK 99504

Arizona   Arizona Dept. of Trans.   Arizona Bureau of
                 206 South 17th Avenue   Geology & Mineral
                 Phoenix, AZ 85007   Technology
                                                     Tucson, AZ 85719

Arkansas   Arkansas State Highway & Trans.   Arkansas Geological
                    Department   Commission
                 P.O. Box 2261   Wardelle Parham
                 Little Rock, AR 72204   Geological Center
                                                     3815 West Roosevelt
                                                     Little Rock, AR 72204

California   California DOT   Calif. Dept. of
                 P.O. Box 1139   Conservation
                 Sacramento, CA 95805   Div. of Mines & Geology
                                                     Resources Bldg.,
                                                       Room 1341
                                                     1416 Ninth Street
                                                     Sacramento, CA 95814

Colorado[1]   Colo. Dept. of Highways   Colorado Geological
                 4201 East Arkansas Ave.   Survey
                 Denver, CO 80222   1313 Sherman St.
                                                       Room 715
                                                     Denver, CO 80203

Connecticut   Connecticut DOT   Dept. of Environental
                 24 Wolcott Hill Road   Protection
                 P.O. Drawer A   State Office Bldg.
                 Wethersfield, CT 06109   Room 561
                                                     Hartford, CT 06115

Delaware   Delaware Department of Trans.   Delaware Geological
                 P.O. Box 778   Survey
                 Highway Administration Center   University of Delaware
                 Dover, DE 19901   101 Penny Hill
                                                     Newark, DE 19711

District of Columbia	Department of Transportation Presidential Building 415 12th Street, N.W. Washington, DC 20004	
Florida	Department of Transportation Hayden Burns Building 605 Suwannee Street Tallahassee, FL 32301	Department of Natural Resources 903 West Tennessee Street Tallahassee, FL 32301
Georgia	Georgia Dept. of Transportation 2 Capitol Square, S.W. Atlanta, GA 30334	Georgia Dept. of Natural Resources 19 Martin Luther King, Jr. Dr. S.W. Atlanta, GA 30334
Hawaii	Hawaii Dept. of Transportation 869 Punchbowl Street Honolulu, HI 96813	Dept. of Land & Natural Resources Div. of Water & Land Development P.O. Box 373 Honolulu, HI 96809
Idaho	Idaho Transportation Department 3311 West State Street P.O. Box 7129 Boise, ID 83707	Idaho Bureau of Mines & Geology Moscow, ID 83843
Illinois	Department of Transportation 2300 South Dirksen Parkway Springfield, IL 62764	Illinois State Geological Survey 121 Natural Resources Building 615 East Peabody Drive Champaign, IL 61820
Indiana	Indiana State Highway Commission State Office Building 100 North Sentate Avenue Indianapolis, IN 46204	Dept. of Natural Resources Geological Survey 611 N. Walnut Grove Bloomington, IN 47401
Iowa	Iowa Dept. of Transportation 800 Lincoln Way Ames, IA 50010	Iowa Geological Survey 123 North Capitol Iowa City, IA 52242
Kansas	Kansas Dept. of Transportation State Office Building Topeka, KS 66612	Kansas Geological Survey Raymond C. Moore Hall 1930 Avenue A, Campus West University of Kansas

Lawrence, KS 66044

Kentucky	Department of Transportation State Office Building High Street Frankfort, KY 40622	Kentucky Geological Survey University Kentucky 311 Breckinridge Hall Lexington, KY 40506
Louisiana	Louisiana Dept. of Highways P.O. Box 44245, Capitol Station Baton Rouge, LA 70804	Louisiana Geological Survey Box G, University Station Baton Rouge, LA 70893
Maine	Maine Dept. of Transportation State House Station 16 Augusta, ME 04330	Maine Geological Survey Ray Building, State House Station 22 Augusta, ME 04330
Maryland	Maryland DOT Balt.-Wash. Int'l Airport P.O. Box 8755 Baltimore, MD 21203	Maryland Geological Survey Merryman Hall John Hopkins University Baltimore, MD 21218
Massachusetts	Massachusetts Department of Public Works P.O. Box 1775 Boston, MA 02105	Dept. of Environmental Quality Engineering Division of Waterways 100 Nashua Street, Room 532 Boston, MA 02114
Michigan	Michigan Dept. of Transportation Transportation Building 425 West Ottawa P.O. Box 30050 Lansing, MI 48909	Michigan Dept. of Natural Resources Geological Survey Division Stevens T. Mason Building Lansing, MI 48909
Minnesota	Minn. Dept. of Transportation State Highway Building St. Paul, MN 55155	Minnesota Geological Survey University of Minnesota 1633 Eustis Street St. Paul, MN 55108
Mississippi	Mississippi State Highway Dept. Woolfolk Office Building P.O. Box 1850 Jackson, MS 39205	Mississippi DNR, Bureau of Geology 2525 North West Street Box 5348 Jackson, MS 39216

Missouri	Missouri State Highway Dept. State Highway Building Jefferson City, MO 65101	Missouri Geological Survey Div. of Geology & Land Survey P.O. Box 250 Rolla, MO 65101
Montana	Montana Dept. of Highways 2701 Prospect Avenue Helena, MT 59601	Montana Bureau of Mines & Geology Montana College of Mineral Science & Technology Butte, MT 59701
Nebraska	Nebraska Dept. of Roads P.O. Box 94759 Lincoln, NE 68509	Conservation & Survey Division University of Nebraska Lincoln, NE 68508
Nevada	Nevada Dept. of Transportation 1263 South Stewart Street Carson City, NV 89701	Nevada Bureau of Mines & Geology University of Nevada Reno, NV 89507
New Jersey	New Jersey DOT 1035 Parkway Avenue Trenton, NJ 08625	New Jersey Geological Survey 1911 Princeton Avenue Trenton, NJ 08625
New Mexico	New Mexico State Highway Dept. 1120 Cerillos Road P.O. Box 1149 Santa Fe, NM 87503	New Mexico State Bureau of Mines & Minerals Resources New Mexico Tech Socorro, NM 87801
New Hampshire	New Hampshire Dept. of Public Works & Economic Development John O. Morton Building 85 Loudon Road, P.O. Box 483 Concord, NH 03301	Dept. of Resources & Economic Dev. James Hall, University of New Hamphire Durham, NH 03824
New York	New York Dept. of Transportation State Campus 1220 Washington Avenue Albany, NY 12232	New York State Geological Survey New York State Education Bldg. Room 973 Albany, NY 12224
North Carolina	North Carolina Dept. of Trans. P.O. Box 25201 Raleigh, NC 27611	Dept. of Natural & Economic Resources Office of Earth

		Resources P.O. Box 27687 Raleigh, NC 27611
North Dakota	North Dakota State Highway Dept. State Highway Building Capitol Grounds Bismarck, ND 58505	North Dakota Geological Survey University Station Grand Forks, ND 58201
Ohio	Ohio Dept. of Transportation 25 South Front Street Columbus, OH 43215	Ohio Dept. of Natural Resources Division of Geological Survey Fountain Square, Building 6 Columbus, OH 43224
Oklahoma	Oklahoma Dept. of Transportation 200 N.E. 21st Street Oklahoma City, OK 73105	Oklahoma Geological Survey University of Oklahoma 430 Van Vleet Oval Room 163 Norman, OK 73019
Oregon	Oregon Dept. of Transportation Room 17, Highway Building Salem, OR 97310	State Dept of Geology & Mineral Industries 1005 State Office Building Portland, OR 97201
Pennsylvania	Pennsylvania Dept. of Trans. Transportation & Safety Building Commonwealth & Forster Streets Harrisburg, PA 17120	Dept. of Environmental Resources Bureau of Topographic & Geo. Survey P.O. Box 2357 Harrisburg, PA 17120
Rhode Island	Rhode Island DOT 368 State Office Building Smith Street Providence, RI 02903	Graduate School of Oceanography University of Rhode Island Kingston, RI 02881
S. Carolina[1]	S. Carolina Dept. of Highways & Public Transportation State Highway Building P.O. Box 191 Columbia, SC 29202	South Carolina Geological Survey Harbison Forest Road Columbia, SC 29210
South Dakota	South Dakota DOT State Highway Building Pierre, SD 57501	South Dakota State Geological Survey Science Center, Univ.

		of S. Dakota Vermillion, SD 57069
Tennessee	Tennessee DOT 105A Highway Building Nashville, TN 37219	Dept. of Conservation Division of Geology G-5 State Office Building Nashville, TN 37219

Texas	Texas Dept. of Highways & Public
	  Transporation
	Highway Building
	Austin, TX 78701
		Bureau of Economic
		  Geology
		University of Texas at
		  Austin
		University Station
		  Box X
		Austin, TX 78712

Utah	Utah Dept. of Transportation
	State Office Building
	Salt Lake City, UT 84101
		Utah Geological &
		  Mineral Survey
		State of Utah-Dept. of
		  Natural Resources
		606 Black Hawk Way
		Salt Lake City, UT
		84108

Vermont	Vermont Agency of Transportation
	State Administration Building
	133 State Street
	Montpelier, VT 05602
		Office of State
		  Geologist
		Agency of Environmental
		  Conservation
		Heritage II, State
		  Office Bldg.
		P.O. Box 79, River
		  Street
		Montpelier, VT 05602

Virginia	Virginia Dept. of Highways &
	  Transportation
	1221 East Broad Street
	Richmond, VA 23219
		Division of Mineral
		  Resources
		Natural Resources
		  Building
		P.O. Box 3667
		Charlottesville, VA
		22903

Washington	Washington DOT
	Highway Administration Building
	Olympia, WA 98504
		Dept. of Natural
		  Resources
		Geology & Earth
		  Resources Div.
		Olympia, WA 98504

West Virginia	West Virginia Dept. of Highways
	1900 Washington Street, East
	Capitol Complex
	Charleston, WV 25305
		West Virginia
		  Geological & Economic
		  Survey
		Mont Chateau Research
		  Center

		Morgantown, WV 26505

Wisconsin[1]	Wisconsin Dept. of Trans.
	P.O. Box 7426
	Madison, WI 53707
		Wisconsin Geological &
		  Natural History
		  Survey
		University of Wisconsin
		1815 University Avenue
		Madison, WI 53706

Wyoming	Wyoming Highway Dept.
	State Highway Office Building
	P.O. Box 1708
	Cheyenne, WY 82001
		Geological Survey of
		  Wyoming
		P.O. Box 3008,
		  University Station
		University of Wyoming
		Laramie, WY 82071

[1]See State and City Mapping, State Cartographer

APPENDIX III - REFERENCES

1. American Railway Engineering Association, "Specifications For Preparation of Maps and Profiles," Chicago, 1962, AREA Manual, Chapter 11, Part 4, p.27.

2. Antill, Paul A. and Gockowski, Jerome A., "Proposal For A National High Altitude Photography Data Base," Proceedings of the 45th Annual Meeting, American Society of Photogrammetry, Falls Church, Va., March, 1979.

3. Doyle, Frederick J., "Digital Terrain Models: an Overview,"Photogrammetric Engineering and Remote Sensing, Dec., 1978, p. 1484.

4. National Ocean Survey, "Catalog of Aeronautical Charts and Related Publications," Rockville, Md., Apr., 1977.

5. Thompson, Morris M., "Maps for America," U.S. Geological Survey, Reston, Va., Second Ed., 1982.

6. U.S. Department of Commerce, Bureau of the Census, "Census Geography," Data Access Description No. 33, Washington, D.C., May, 1979.

7. U.S. Geological Survey, NCIC, "Digital Terrain Tapes, User Guide, Second Edition," Reston, Va., n.d.

8. U.S. Geological Survey, "Nature To Be Commanded," U.S. Geological Survey Professional Paper 950, Reston, Va., 1978

9. U.S. Geological Survey, "Yearbook, Fiscal Year 1978," Reston, Va., 1979, p.62.

APPENDIX A. BIBLIOGRAPHY

Some useful publications on subjects related to surveying and mapping which complement the content of the special publication are listed. Those which do not relate to a specific chapter are under the heading, General. Those which relate more directly to the content of a chapter are listed under the headings: Map Accuracy, Map Content and Symbols, and Map Sources.

General

Bouchard, H., "Surveying," revised by F.H. Moffitt, International Textbook Co., Scranton, PA, 1965.

Coastal Mapping Handbook, M.Y. Ellis, ed., National Ocean Survey and U.S. Geological Survey, United States Government Printing Office, Washington, D.C., 1978.

Davis, R.E., Foote, F.S., and Kelly, J.W., "Surveying Theory and Practice," McGraw-Hill Book Co., New York, NY, 1966.

Definitions of Surveying and Associated Terms, ASCE Manual No. 34, by a Joint Committee of the American Society of Civil Engineers and the American Congress on Surveying and Mapping, ASCE, 1978.

Deitz, C.H., and Adams, O.S., "Elements of Map Projection," U.S. Coast and Geodetic Survey Spec. Pub. 68, United States Government Printing Office, Washington, D.C., 1944.

Glossary of Mapping, Charting, and Geodetic Terms, Defense Mapping Agency Topographic Center, Washington, D.C., 1973.

Hickerson, T.F., "Route Surveys and Design," McGraw-Hill Book Co., New York, NY, 1959.

Hydrographic Manual, M.J. Umbach, ed., NOAA/National Ocean Survey, Rockville, MD, 1960.

Kissam, P., "Surveying Instruments and Methods," McGraw-Hill Book Co., New York, NY, 1947.

Kissam, P., "Surveying for Civil Engineers," McGraw-Hill Book Co., New York, NY, 1956.

Manual of Color Aerial Photography, J.T. Smith, ed., American Society of Photogrammetry, Falls Church, VA, 1980.

Manual of Photogrammetry, C.C. Slama, 4th ed., American Society of Photogrammetry, Falls Church, VA, 1980.

Manual of Remote Sensing, R.N. Colwell, 2nd ed., American Society of Photogrammetry, Falls Church, VA, 1983.

Manual of Surveying Instructions, Technical Bulletin 6, Bureau of Land Management, Washington, D.C., 1973.

Moffitt, F.H., "Photogrammetry," 2nd edition, International Textbook Co., Scranton, PA, 1967.

Report on Highway and Bridge Surveys, Manual No. 44, Committee on Engineering Surveying, Surveying and Mapping Division, American Society of Civil Engineers, New York, NY, 1962.

Specifications for Preparation of Maps and Profiles, AREA Manual, American Railway Engineering Association, Chicago, IL, 1962.

Technical Procedure for City Surveys, Manual No. 10, Committee on City Surveys, Surveying and Mapping Division, American Society of Civil Engineers, New York, NY, 1963.

Topographic Maps, U.S. Geological Survey, United States Government Printing Office, Washington, D.C., 1981.

Urban Planning Guide, Manual No. 49, Committee on Review of Urban Planning Guide, Urban Planning and Development Division, American Society of Civil Engineers, New York, NY, 1969.

Wilford, J.N., "The Map Makers," Alfred A. Knopf, Inc., New York, NY, 1981.

Wolf, P.R., "Elements of Photogrammetry," McGraw-Hill Book Co., New York, NY, 1974.

Map Accuracy

Burington, R.S., and May, D.C., "Handbook of Probability and Statistics with Tables," McGraw-Hill Book Co., New York, NY, 1970.

Classification, Standards of Accuracy and General Specifications of Geodetic Control Surveys, NOAA/National Ocean Survey, Rockville, MD, 1974.

Mikhail, E.M., and Gracie, G., "Analysis and Adjustment of Survey Measurements," VanNostrand Reinhold Co., New York, NY, 1981.

Principles of Error Theory and Cartographic Applications, Technical Report No. 96, Aeronautical Chart and Information Center, St. Louis, MO, 1962.

Reference Guide Outline, Specifications for Aerial Surveys and Mapping by Photogrammetric Methods for Highways, Photogrammetry for Highways Committee, American Society of Photogrammetry, Federal Highway Administration, Washington, D.C., 1968.

Specifications to Support Classification, Standards of Accuracy, and

General Specifications of Geodetic Control Surveys, NOAA/National Ocean
Survey, Rockville, MD, 1975.

Map Content and Symbols

Abbreviations to be Used on Drawings and in Text, (ANSI Y1.1), American
National Standards Institute, American Society of Mechanical Engineers,
New York, NY, 1972.

ANSI Series Y32.xx to include Z32.2.3, Z32.2.4, Z32.2.6, American
National Standards Institute, American Society of Mechanical Engineers,
New York, NY.

Chart No. 1, Nautical Chart Symbols and Abbreviations, National Ocean
Survey and Defense Mapping Agency, Department of Defense, Washington,
D.C., 1979.

Dreyfuss, H., "Symbol Sourcebook," McGraw-Hill Book Co., New York, NY,
1972.

Glossaries of BLM Surveying and Mapping Terms, Cadastral Survey Training
Staff, Denver Service Center, Bureau of Land Management, Denver Co.,
1980.

Keates, J.S., "Cartographic Design and Production," John Wiley and Sons,
New York, NY, 1973.

Raisz, E., "Principles of Cartography," McGraw-Hill Book Co., New York,
NY, 1962.

Report on Uniform Map Symbols, Parts I and II, Subcommittee on Uniform
Map Symbols of the Committee on Highway Planning, The American
Association of State Highway Officials, Washington, D.C., 1962.

Robinson, A.H., Sale, R.D., and Morrison, J.L., "Elements of
Cartography," John Wiley and Sons, New York, NY, 1978.

Visual Aeronautical Chart Symbols, NOAA/National Ocean Survey and
Defense Mapping Agency, Department of Commerce, Washington, D.C., 1974.

Wattles, G.H., "Surveying Drafting," Gurden H. Wattles Publications,
Orange, CA, 1977.

Wilkins, E.B., "Maps for Planning," Public Administration Service,
Chicago, IL, 1948.

Map Sources

Catalog of Aeronautical Charts and Related Publications, NOAA/National
Ocean Survey, Department of Commerce, Washington, D.C., 1981.

Census Geography, Data Access Description No. 33, Bureau of the Census, Washington, D.C., 1979.

Dates of Latest Editions, Nautical Charts & Misc. Maps, Quarterly, NOAA/National Ocean Survey, Rockville, MD, 1982.

Digital Terrain Tapes, User Guide, 2nd ed., National Cartographic Information Center, U.S. Geological Survey, Reston, VA.

Indexes to Topographic Maps, National Cartographic Information Center, U.S. Geological Survey, Reston, VA.

Nature to be Commanded, Professional Paper 950, U.S. Geological Survey, Reston, VA. 1978.

Nautical Chart Catalogs 1 through 5, NOAA/National Ocean Survey, Rockville, MD, 1982. (Annual)

Thompson, M.M., "Maps for America," U.S. Geological Survey, Reston, VA., 2nd edition, 1982.